HELP YOUR CHILD
WITH MATHS

Help your child
with
MATHS

Edited by Angela Walsh

BBC Books

BBC Books wish to thank the authors: Andree Brough, Jackie Cook,
Mary Salvage, Christine Thomas and Angela Walsh for the co-operation, dedication
and hard work that they have contributed to this publication.

Published to accompany a series of television programmes
in consultation with the BBC Continuing Education Advisory Council
and the PrIME Project, Homerton College, Cambridge.

Cover montage
Anni Axworthy
Illustrations and diagrams
Richard Geiger

Published by BBC Books,
a division of BBC Enterprises Limited,
Woodlands, 80 Wood Lane, London W12 0TT
First published 1988

© The authors 1988

Reprinted 1989

ISBN 0 563 21444 9

Set in 10 on 12 point Univers Light
Typeset in Great Britain by Ace Filmsetting Ltd, Frome, Somerset
Printed and bound in Great Britain by Richard Clay Ltd, Bungay, Suffolk

Contents

Introduction

Many parents want to help their children to learn maths but they often feel that the maths taught in schools today is different from the type of maths they learned themselves. This sometimes makes it difficult for parents to know just what to do.

This book has been written by a group of teachers who for some time have been working with parents on mathematical activities for children in the four to eleven age range. A wide selection of activities is offered in the book for you to use with your child. These activities have been designed to support and develop your child's education in maths.

Why has maths changed?
Many parents want to know more about the developments that have taken place in maths. Why has there been a change of emphasis towards more practical tasks, towards the process of making maths meaningful, towards encouraging children to solve problems and talk about the maths that they are doing? Is it a good thing?

There are several important reasons for these developments and all of them have support at the highest levels.

The first reason for changing the approach to maths involves how people define maths. When maths is mentioned most people think of numbers and addition, subtraction, multiplication and division. But this is only a small part of maths. Maths is also concerned with building up ideas about relationships between things, with recognising and developing patterns and about building structures. These involve not only number but also geometry, measurement, logic, algebra, statistics and probability, to name just a few. And all of this is available in some form to children, provided the right kind of activities are available. By offering a

7

much broader view of maths we will be helping children to see its many uses and applications, and to recognise the part it plays in life.

A second reason for shifting the emphasis in maths away from 'sums' is because technology is increasingly important in the society in which your children are growing up. We take machines like video recorders and microwaves for granted and enjoy having them because they remove unnecessary hard work. This is also true of computers and calculators. We need to encourage children to develop the skills which are essential for life in a rapidly changing technological age. These are skills which imply a need for people who can work things out for themselves, develop meaning and understanding for the things they do, feel at home with technological tools like computers and calculators and use them to best advantage.

A third reason is related to evidence that shows that large numbers of adults, perhaps even you, find applying maths very difficult. Many adults also say that they feel insecure and lacking in their understanding of maths, despite having spent many years learning it at school. Much of their time was spent in practising 'sums' but little time was used to show how to use this knowledge to solve real problems. Unfortunately, this means that many parents have a negative attitude towards maths and see it as hard, boring and only for 'brainy people'. Many adults are happy to admit 'I was never any good at maths.' This image of maths doesn't help young learners. Maths should not be left just to the few and the new approaches are designed to help *all* children become maths thinkers able to use their mathematics and to develop a positive feeling towards it.

What does this book offer?
Of course it's still important for children to learn how to count and to do 'sums'. Attention also needs to be paid, however, to how these and other mathematical *skills* can help develop an understanding of mathematical *ideas*. Maths will then be of real value to your child in this ever-changing world in which we live.

This book has been designed to help you develop this understanding of mathematical ideas. The following have been taken into full consideration:

Support for children in developing the early skills and concepts, the 'building blocks' of maths.

How these will enable children to meet their maths needs in later learning and in adult life.

How you can best help your children develop their maths skills.

The activities

We have provided a series of activities which we feel will offer your children the opportunity to experience maths in an enjoyable way.

They are designed to enable parents and children to 'play' constructively, and to use their problem-solving skills within maths. The activities are divided into two main sections, *In the Home* and *Outdoors*. Each of these is subdivided to offer activities suitable for *four- to eight-year-olds* and *seven- to eleven-year-olds*. Your child, however, may well enjoy some activities from either set – it really depends on their interest and previous experience. The best bet is to try some and see.

Maths in the home

The home is where a child's maths education begins. It also provides a wealth of opportunities for maths activities and games. Here, both you and your child can be together in a relaxed setting which is relevant to the child. It also offers you a chance to share maths with your child without too much extra effort!

Maths outdoors

The outdoor activities provide opportunities for children to continue developing maths in the outside world. The activities cover a wide range of ideas which can mostly be carried out as part and parcel of everyday events. If, for example, as you shop you chat about money, time or shape, your child will be using maths and seeing it being used.

On the whole activities have been chosen because they require little by way of unusual materials or preparation. Reading them through first, however, is likely to prove helpful.

Why should I become involved?

Evidence from other parents suggests that helping your child with maths can bring the following benefits:

- you learn more about your child;
- you find that talking about school and school work comes more easily;
- you reinforce a positive attitude to maths because you're showing you value it;
- you encourage your child to ask more questions and become more mathematically aware;
- you help them to get a thorough grounding in the subject and experience with them the pleasure of learning;
- you encourage them to 'talk maths' and to develop their language skills;

9

- you help them to develop 'mathematical thinking', involving problem-solving and investigation;
- you learn alongside your child;
- you share many interesting and enjoyable experiences.

Lastly, but not least, you can enjoy doing some maths yourself.

To make the experience as rewarding as possible, here are some points to bear in mind when working with your child:

1 Talk and listen to your child as much as possible during any activity.
2 Encourage your child so as to get enjoyment and excitement from doing the activity.
3 Encourage your child to do and to understand the maths in the activity.
4 Try to help your child to develop powers of logical thinking,
5 Support your child in developing an awareness of mathematics in the everyday world.
6 Help and encourage your child to become a problem-solver. Don't try to impose methods or solutions on the child. Help the child to use the ones with which they feel happy and guide them to discover new ways. Remember there are often many solutions to any problem.

Each of the sections contains activities on a variety of mathematical topics, such as symmetry, shape, number and pattern. This is to provide a wide-ranging view of mathematics. The topics covered are listed at the start of each activity. An index of topics is also provided to help you find all the activities on a particular topic should you so wish.

In the home

4- to 8-year-olds

JIGSAW PUZZLES

Logical thinking   Shape/size   Area

You will need an old birthday card, Christmas card or picture.

On the back of the picture draw some lines from edge to edge. They can be curved or straight, or both. Now cut out your jigsaw puzzle pieces. Take the pieces and try to put the picture together again, then give it to someone else to do.

Now find a plain piece of rectangular card. Draw four or five lines in pencil across the card. Cut out the shapes and jumble them up.

Now try to fit the shapes together. It's not as easy as it seems, and of course the more lines you draw, the more difficult it becomes.

STRINGY PROBLEMS

Length/distance   Estimating/approximating

Cut some lengths of string. Lay them on the floor in different shapes.

Guess which piece is the longest and which piece is shortest. Were you right?

Draw the numbers 1–5 on large sheets of paper.
Cut five lengths of string to lay over these numbers.
Which number will need the longest piece of string?
Can you guess the order from longest to shortest before you cut the string?

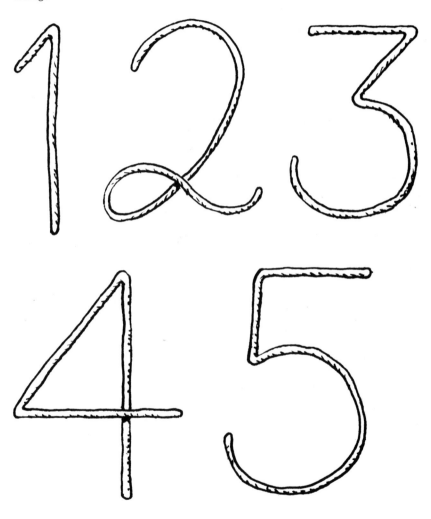

Now do the same thing with the letters in your name.

CORNFLAKE DELIGHT

Language/talking Capacity/volume Weight
Number patterns/ideas

Recipe
1 oz (25 g) butter
2 tablespoons golden syrup
1 tablespoon cocoa powder
2 tablespoons granulated sugar
7 tablespoons cornflakes

1 Count out 20 paper cases and put them on to a wooden board.
2 Put butter into a saucepan and add syrup. Leave over a low heat until butter melts. Then stir in cocoa.
3 Take saucepan away from heat and add sugar. Sprinkle in cornflakes. Stir quickly until cornflakes are coated with cocoa mixture.
4 Using two teaspoons, pile into paper cases. Leave to set.

Now you have made your Cornflake Delights, will there be enough for each of your family to have an equal share?

THE KITCHEN CUPBOARD

Language/talking Shape/size

Look at a box of cornflakes. (The mathematical name for this shape is a cuboid.)
If you drew a face on each side of the box, how many faces would you draw?
How many edges has this box?
How many corners?
Are there any other things in your kitchen that have the shape of a cuboid?
What is the largest cuboid you can see? What is the smallest?

Look at an OXO cube. Why is this shape different from a cuboid?

Look at a tin of baked beans. (The mathematical name for this three-dimensional (3-D) shape is a cylinder.)
Two of the surfaces are circular. What shape is the other face? You may need to put a piece of paper round the tin to help you find out.
Are all the tins in your cupboard cylindrical?
What is the smallest cylinder you can see? What does it hold?

Are there any other interesting solid shapes in your kitchen?

Before you throw away a box, open it out carefully. This is the net of the box.

Look at several nets of boxes. You may now like to design your own box or container.

STAIRS

Language/talking   Number patterns/ideas

How many stairs are there in your house?
Can you put one toy on each step?
How many toys will you need?
Can all your family stand on one step each?
Are any steps empty?
How many?

STORIES AND RHYMES

Language/talking   Investigating

It is important to develop a child's ability to describe and make comparisons. The comparisons that mathematicians are particularly interested in are those of number, size, shape and position, and many of these ideas can be developed quite naturally while telling stories and rhymes.

A story always has a beginning, a middle and an end. After reading a story it can be useful to ask a child to tell the story in their own words. Being able to relate a sequence of events is a skill that is as important for mathematics as for language. Stories can also provide a focus for some practical and problem-solving situations. Outlined below are some suggestions for follow-up activities for a favourite story – 'The Three Bears'. Make sure you have all the necessary materials before you begin.

Design a bed for your teddy. Make a pillow and bedclothes (cereal boxes and scraps of material will be needed). Use some plasticine to make models of the three bears. Make chairs for them using Lego or cardboard boxes.

Find out how porridge is made. How many cupfuls of porridge oats and how many cupfuls of water do you need to make porridge for three people?

DOMINOES

Addition

Variations of domino games are an ideal way to practise number bonds (two numbers which add up to a given total). All you need is a set of dominoes.

Make Six (a game for 2–4 players)
Each player takes five dominoes and looks at them. One of the remaining dominoes is turned up in the middle, the rest remain face down to one side in the 'bank'.
The first player must place a domino on one end of the one already placed to make a total of six spots.
Each player takes it in turns. If they can't go they take one from the 'bank' and either play that domino or add it to their pile.
The first player to put down all their dominoes wins. The other players total the spots on the dominoes they have left. After several games, the player with the lowest score is the winner.

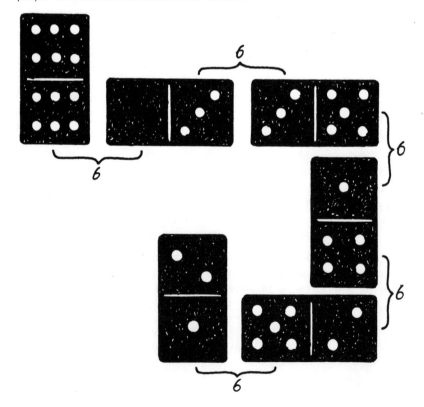

I SPY

Language/talking   Addition   Multiplication

A bird has two legs, a cat has four legs, an insect has six legs, a spider
has eight legs.
I spy with my little eye – ten legs. Do I see a dog and a fly or four birds
or what?

Make up some of your own number stories and try them out on a friend.
Try and find some pictures of animals and insects or birds and use these
to spy with your little eye.

BUTTON BOX

Sorting/classifying   Shape/size   Estimating/approximating

You will need a collection of buttons, pieces of card of various sizes,
pencil, a large tray, empty yoghurt or margarine tubs, scissors.

Tip the buttons onto a tray so that they can be spaced and easy to see.
Put the tubs around the edge and let the children sort them out.

They can be asked to sort them in various ways:
    colour
    plain or patterned
    clear or opaque
    shiny or dull
    number of holes
    large or small
    matching sets

Once they have been sorted into matching sizes or sets, ask the children
to find or cut out a piece of card that one of their sets will fit on to. The
set or sets can then be stitched on (only a couple of stitches for each
button are needed). Try to get several sets of buttons of the same
number, e.g. six.
Ask how many ways sets of six can be arranged on a card.
Can all sets be arranged in different ways?
Now try to sort the sets according to how they might be used. Ask:
    Which would be used for shirts or blouses?
    Which would be used for a baby's clothes?
    Which would be suitable for an outdoor coat or jacket?

17

FUNNY STORIES

Language/talking   Addition   Multiplication

Make up a number story. For example:
 There were two cars on the road.
 In each car there were two people.
 Each person had two bags.
 In each bag were two boxes.

How many boxes were there altogether?
Is there a way of drawing or writing this down to make the calculation easier?

Can you use your calculator to find the total?

Make up some of your own number stories and try them out on a friend.

CLOWNS

Number patterns/ideas   Addition   Shape/size

This is a game for any number of players. You will need paper, pencil and dice.

Players each begin with a circle for the clown's face and then take it in turns to throw the dice.

Throw a 6 for a hat,
       5 for a ruff,
       4 for an eye,
       3 for a nose,
       2 for a mouth,
       1 for an eyebrow.

Each player draws their clown's features as they go along. The first one to complete the clown is the winner.

Variations

The odd or even clown
The game can be extended by using two dice and scoring with only odd or even numbers.

In the even game		*In the odd game*
Throw a 12	for a hat	Throw a 11
10	for an eye	9
8	for an eyebrow	7
6	for a nose	5
4	for a mouth	3
2	for a ruff	1

The shape clown
In this game players need to throw:
  a 6 for a large circle – the face
  a 5 for a small circle – the pom-pom
  a 4 for a large triangle – the hat
  a 3 for a small triangle – the nose
  a 2 for a large rectangle – the mouth
  a 1 for a small rectangle – an eye.

The pieces should be cut out ready for the players to pick up.

DIGITAL NUMBERS

Number patterns/ideas

How many displays of digital numbers can you find in your home?

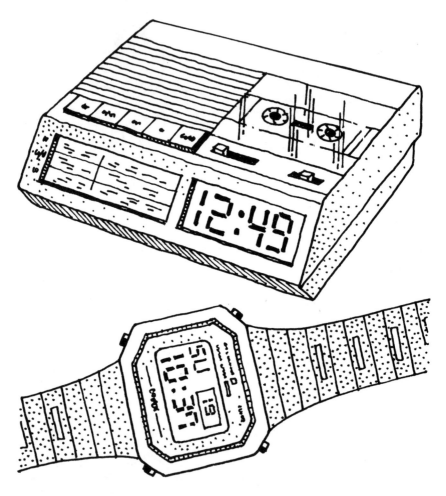

Look at a calculator – it displays numbers in a digital form. Try to make some digital numbers using matchsticks.

How many matchsticks did you use for each number?

How are digital numbers different to the ones we write?

SPOT THE DELIBERATE MISTAKE

Addition Subtraction Multiplication Division Money Shape/size Fractions

Make up a story and put in deliberate mathematical mistakes. Children have to spot the mistake and correct it. For example:

Maria and her three friends were on the beach. The five of them were building a dam to stop the tide coming in. 'If we all make two sand castles with our buckets, we can build a wall seven castles wide,' she said.

'I haven't got a bucket,' said Sara, 'but my Dad gave me 50p to spend. How much are they?'

'I think they are 45p,' said Taz, 'so you will have 10p change to buy us all a 3p ice cream.'

OPEN THE BOX

Estimating/approximating Mental calculation Subtraction Multiplication

A game for at least one adult and any number of children. (Preparation is needed for this game. The children can help and a 'junk box' will be invaluable.)

Between five and eight packages are needed, graded in size so that each will fit inside another, like a Russian doll. For example, a jewellery box inside a matchbox, inside a larger matchbox, inside a teapacket, inside a washed litre juice carton, inside a cereal packet, etc.

Inside the smallest one is a 'treat' of some kind. Then the boxes are placed inside one another in order.

The game is then played like pass the parcel. The difference is that when the music stops, the player is asked a maths question by the adult before they can remove the next box. The one who removes the last box gets the 'treat'.

COUNTERS

Logical thinking Subtraction

A game of strategy for two players. You will need 20 counters or buttons.

The counters are placed on the table.
At their turn each player must remove at least one but no more than three counters.
The loser is the player who has to remove the last counter.

SHELVE IT

Arranging Patterns

Put three toys, side by side, on a shelf.
How many different ways can the three toys be arranged?

Next time, try it with four toys, but first try to guess how many different arrangements there will be.

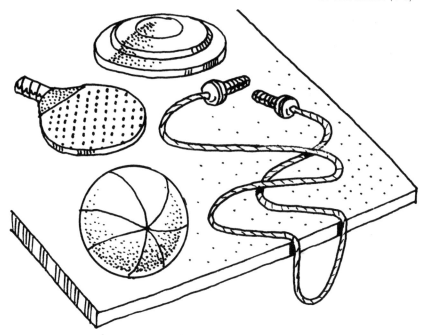

The number of toys can be increased. Is there a pattern in the number of possible arrangements?

HUNT THE 'SLIPPER'

Sorting/classifying   Language/talking

Collect five or six pairs of shoes and sort them into order according to length, shortest first up to longest. This isn't easy and plenty of help comparing lengths will be needed. It helps if the shoes are placed against a straight edge. This makes it easier to see if one shoe is longer than another.

Once sorted, one of each pair is taken and hidden by one person. The object of the game is for the shoes to be returned in order, shortest first, etc. Positioning clues may be given, like:

It's under a . . .
on top of a . . .
next to a . . .
beneath a . . .
inside a . . .

A variation of this game is to hunt the 'slipper' according to each person's height.

23

JUMBLE SHOES

Sorting/classifying   Pairing   Angle/direction   Language/talking

You will need a collection of footwear in a large box, paper and pencil.

First sort the footwear into pairs. Make sure that left and right are placed correctly.

Find the longest pair and the shortest pair and draw round them. Choose another two pairs, draw round them and place the drawings in order of length between the drawings of the first two pairs. Do the same for the remaining pairs of shoes.

You can now make a simple picture graph of shoe lengths.

> Try working out whose shoes were the shortest.
> Whose were the longest?
> Is there a shoe longer/shorter than . . . ? (Point to a shoe or name it, e.g. Dad's slipper.)

Variations

1 The collection could be limited only to slippers, trainers, wellington boots, or one person's footwear.
2 The footwear could be sorted according to owner's height. A piece of paper attached to the wall at ground level could be marked to make the comparison of height easier.

TOWERS

Language/talking   Length/distance   Shape/size

Build a tower with bricks, Lego, paper cups or yoghurt pots.
How many different ways can you build a tower?
Which tower uses the most/fewest bricks?
Can you build a very tall tower and very strong tower?
What is the difference between the two towers?

Build a wall like this:

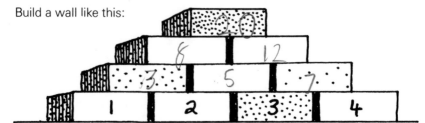

It has four bricks at the bottom. How many rows are there?
How many bricks do you need altogether?
Now build a wall with six bricks at the bottom. Before you start guess
how many rows there will be and how many bricks you will need.
What other kinds of wall can you build?

THINK OF A NUMBER

Addition Subtraction Multiplication Division Shape/size
Money Logical thinking Language/talking

There are numerous ways of playing this 'I'm thinking of a number' game.

At its simplest level you can say 'I'm thinking of a number between 0 and
10.' As the children try to guess the number, answer with 'It's more than
that' or 'It's less than that.' At first they will just make random guesses but
gradually they will develop strategies to guess with fewer tries. You can
encourage them to develop more organised strategies, e.g. 'Is it more
than 10?', 'Is it less than 15?'

To help with specific ideas children may be meeting at school, try:
I'm thinking of an even/odd number between . . . and . . .
I'm thinking of a number which is 2 less than/5 more than . . ., etc.
I'm thinking of a number in the 3× table between 16 and 20.
I'm thinking of a number 1 more than/1 less than 100/1000.

Children could make some up for you and the rest of the family.

Variations
You can play this game with:
Mathematical shapes – I'm thinking of a shape with 3 equal sides, etc.
Money – I'm thinking of an amount 7p more than 19p, etc.

The aim of these games is to guess the number correctly with the least
number of questions. Young children may find logical thinking like this
very difficult. It is well worth helping them to develop it.

COVER UP

Addition   Multiplication

Any number of people can play this game. You will need 2 dice, some counters, and paper.

Each player has a board on which the numbers 2–12 are written. Each player takes a turn to throw both dice. The numbers thrown are added and the player covers that number on their board with a counter. The winner is the first player to cover six numbers on their board.

With older children, this game can also be used for multiplication practice using the numbers from 1×1 to 6×6.

WRAPPING PAPER (sorting games)

Sorting/classifying   Counting

You will need wrapping paper with repeated patterns. Paste it on to card and cut out each repeat. When choosing paper to cut up in this way, look for patterns which have some variation. For example, the figure or shape repeated in different colours, or doing different activities, or holding different objects.

Once made, the cards can be used for matching games, playing snap or be sorted into sets of the child's choosing. Laid face down on the table, they can be turned over to find pairs, the odd one out or matching sets. If a duplicate set is made players can have a set each. Each player has the set face up in front of them. One player chooses a card, without touching it in any way. The other player has to ask questions to decide which card was chosen. E.g. Has the teddy bear a striped balloon? Is the teddy bear holding a ball?

THE COORDINATE GAME
(for 3–6 players)

Coordinates

You will need two sheets of identical wrapping paper which has a variety of different animals, vehicles, objects or characters on it.

Divide both sheets into sections to make 32 rectangles (see diagram). Cut one sheet into the 32 pieces and label the other as a grid. Letters A to D along the bottom edge, numbers 1 to 8 along the side.

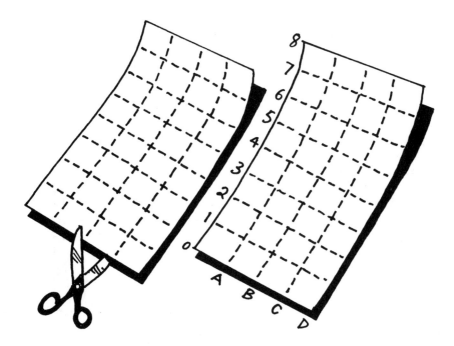

The 32 rectangles are laid face up on the grid. Each player has to ask for a rectangular card, naming what is on the picture and its coordinates. E.g. 'May I have the elephant card in D,6?'

If the coordinates are correct, the player takes the card. The player with the most cards wins.

The game can be made more difficult by changing the size of the grid and by relabelling the bottom edge with numbers. This would make the elephant card in position 4,6.

STEP THROUGH THE WOOD

Language/talking Subtraction

For this game you will need 2 dice and 2 counters. Take turns.

1 Throw the dice.
2 Find the difference between the two numbers, e.g. 3 and 5 – the
difference is 2.
3 If the difference is 2 move your counter forward 2, if it is 4 move your
counter forward 4, etc.
4 The winner is the first person to the treasure.

CARD-TALK

Language/talking Addition Subtraction Multiplication
Sorting/classifying Shape/size Area

Have you ever noticed how young children are fascinated by the cards they receive at Christmas or their birthday? They often like to keep them in a safe place and bring them out from time to time. Here are a few suggestions to encourage 'talking points':

How could we sort the cards?
Which card is the biggest?
Is the widest card also the tallest?
Are any cards the same size?
Put the cards side by side. How far will they reach?
How many cards altogether? How many if I hide two?

You may like to develop this activity further by encouraging your child to make a collection of all types of cards.

FUNNY PICTURES

Shape/size

Draw a picture using squares only, or triangles only, or a combination of both.

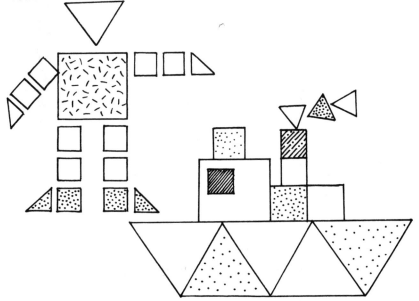

29

TOP HATS – a simple sponge recipe

Weight Estimating/approximating Time Collecting/using information

You will need scales, bowl, spoons, bun tins and paper cases, jam, two eggs, flour, sugar and margarine.

No accurate weighing is necessary. Place the two eggs on the scales and balance them, in turn, with the flour, sugar and margarine.
Put the sugar and margarine into the bowl and mix them together. Break the two eggs and add them to the mixture, along with the flour. Mix all the ingredients thoroughly.
The oven needs to be set to 375°F or 190°C or gas mark 5. Ask the child to set the dial and switch on the oven.

Before the sponge mixture is put into the paper cases ask questions like:
  What size spoon do we need to put the mixture into the cases?
  How many little buns will the mixture make?
  How many paper cases shall we put in the tins?

As the mixture is put into the cases remember to ask:
  Have we enough mixture to fill these cases?
  Do we need to get out some more cases, or another bun tin?

The sponge cakes will take 20 minutes to cook. Once the tins are in the oven ask:
  Can you set the timer for 20 minutes?
  What time by the clock will the cakes be ready?
  If we check them five minutes earlier, what time will that be?
  Can we clear everything away in 20 minutes?
  Shall we get the cooling tray ready?
  What size spoon do we need to put jam in the cakes?

Once the cakes are out of the oven, check the time and ask:
  Were they ready on time, earlier or did we cook them longer?
  How long will they take to cool?
  What time will we put the jam in?

Once the cakes are cool slice off the top, put in a spoonful of jam and put the 'top hat' back on.

RUBBING COINS

Language/talking   Money   Shape/size

You will need thin paper, a pencil or crayons and some coins.

To make an effective rubbing, place your coin/s under thin paper and
gently rub the paper over it/them with a crayon or pencil.
Design your own picture of a person by using coins to the value of £1.
Think of a title for your picture, e.g. Mr or Mrs Poundsworth.
Some titles for you to think about:

    The Tuppenny Tortoise
    The Trembling 10p Tower
    Twenty-pence Tots

Handling coins and rubbing them helps children to recognise coins in
daily use and to learn about their value.

FUN WITH MIRRORS

Symmetry   Patterns   Number patterns/ideas

Make a drawing of a simple shape. What new shapes can you make by
putting a mirror in different positions on your drawing?
Try to draw some of these new shapes.
Now put a mirror across the dotted lines below.

Is the number sentence true or false?

Now make up some of your own and try them out on a friend.
Remember, you will only be able to use numerals that are symmetrical.

EMPTY CARTONS

Estimating/approximating

Matchboxes

The smallest size matchbox holds on average 43 matches. Using an empty matchbox ask if it can be filled with 43 or more different things. For example, one grain of rice, one nail, one button, etc.

Shoeboxes

A shoebox holds two shoes comfortably. Can the children find two other things that would fit as comfortably? The pair do not need to be identical. For example, it could be two different boxes or two different toys.

Tidying the junk box

You will need a large cardboard carton and a collection of empty packets. (Try not to throw out packaging – keep it in a junk box instead.)

Ask such questions as:
    Can all the packets be packed away without any of them sticking out above the carton top?
    Is there more than one way to do it?
    Is the carton too large for the empty packets?
    Could all the empty packets fit into a smaller box?

TAKE A LINE FOR A WALK

Logical thinking

Without lifting your pencil from the paper, doodle and end up with something like this:

With a ruler you might end up with something like this:

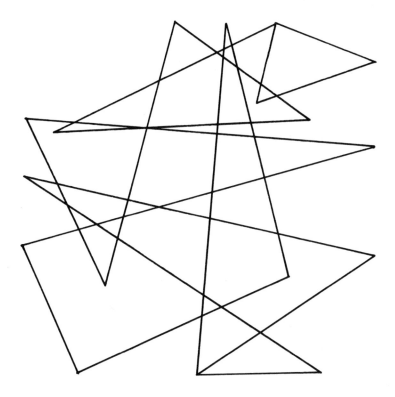

Using no more than four colours, fill in the pattern so that no two neighbouring parts are the same colour. It takes some planning!

TIME DEVICES

Time Sorting/classifying Language/talking

Egg-timers, watches, clocks, oven-timers . . . these are all instruments that measure time.

Find out how many time devices there are in your home. Use one of them to find out how long it takes you to get dressed in the morning.
Time yourself each morning for a week. Did your performance improve?
Could *you* make a simple time device? Here is one idea. Use a plastic bottle with a hole in. Fill it with sand, salt or water.
Other materials which might be useful are a candle, cardboard, cardboard rolls, glue and a marble.

EATING OUT

Money Addition Subtraction

Make a simple menu and price list. The number of items and prices can be adjusted to suit the children. This one is very simple.

The Nutmeg	
Toast	4p
Bread and Butter	3p
Cake	5p
Boiled Egg	7p
Jam	1p
Beans	6p
Lemonade	8p
Milk Shake	9p

Make up little stories and ask the children to answer the questions using the menu. For example:

John had a boiled egg and a slice of toast for breakfast. How much did he spend?

Tracy spent 8p on two things. What did she buy?

Peter spent 16p on three things for his lunch. What did he have?

What could you buy to eat and drink if you had 20p?

This idea can be adapted in several ways. You could use pictures of toys and price each one differently, or items of clothing and ask the cost of various outfits. Stories that interest and involve the children are much better than just asking straightforward 'sums'. Try to ask questions that have more than one answer.

PAPER FOLDING

Patterns *Fractions*

Many traditional paper-folding exercises, for example, to make hats, boats or aeroplanes, are mathematical; they involve halving and quartering by folding.
Start with sheets of paper of different sizes and thicknesses.
Fold in half, in half again and so on.
At which point does it become impossible to fold any more?

MAKE TEN

Addition *Subtraction* *Multiplication* *Division*

A calculator might be useful for this activity.

Have you ever thought of how many sums you could make up with the answer 10?

Here are some:

$6 + 4$

$3 \times 3 + 1$

$19 - 9$

$40 \div 4$

Your turn now – the possibilities are endless!

MEAL TIMES

Arranging *Patterns*

Each time you sit down to a meal, try to find a different seating plan for the family around the table.

When you have exhausted all the possible arrangements, introduce a rule. For example, Mum must sit nearest the kitchen door or Dad must sit at one end of the table. How many arrangements are there now? Supposing a friend comes to tea, how many arrangements will be possible?

Do you have to work this out at the table or could you use play people, buttons, counters or paper and pencil?

FUN WITH THREE DIGITS

Logical thinking   Number patterns/ideas

How many different numbers can be made from any three digits? For
example, choose 123 and make 132, 213, 231, etc.
Which is the biggest? Which is the smallest?
Try it with other sets of three digits.

Try a similar idea using Lego cubes.
How many 'different' towers can you build using a red, a white and a
blue?
What happens if I use a yellow cube as well?

ONE MINUTE

Time

You will need a clock or watch with a 'minute' hand.

Close your eyes and keep them closed till you think one minute has
passed.
Can you chew a carrot or a piece of fruit for one minute?
Stand still, then jump up when you think one minute has passed.
Count the number of times you can clap your hands or jump in the air in
one minute.

PAIRS

Logical thinking   Predicting

Mix up three pairs of socks and put each sock into a drawer individually.

How many socks must be taken out (without looking) to be sure you have
a pair?
Repeat the game with four pairs. Before you start, guess how many you
will need to take out. Then try it.
How about five pairs? Is there a connection between the number of pairs
of socks and how many you must take out to be sure of a pair?
Without actually doing it, can you say with certainty how many you
would need to take out for ten pairs of socks?

CALCULATOR STRAWS

Sorting/classifying   Calculator use

Cut some straws in half or into three equal lengths.
Make 'calculator numbers' with the straws by copying a calculator
display. Make all the numbers from 0 to 9.

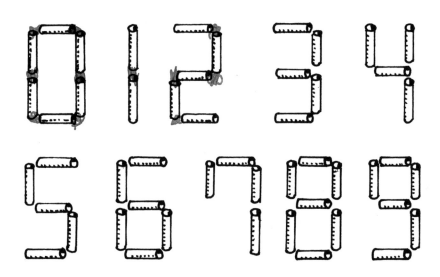

Ask questions like:
  How many straws do you need to make the number 6?
  Which number needs the most/the least straws? How many?
  If you had six straws which numbers could you make?
  Which numbers are made with the same number of straws as their
  own number?
  How many straws do you need to make the number 28?
  Which two-figure numbers could you make with ten straws?

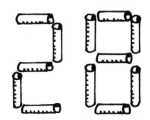

GROUPING

Number patterns/ideas   Patterns

Start with an unknown small number of buttons, counters or marbles.
How large a number can you recognise without counting? Count them.

Is it easier to count them when they are grouped in twos or fives? Try it.
Now how large a number of buttons can you recognise without counting?
How?
Have you got an odd or an even number of buttons?
How can you 'see' the difference between odds and evens?
Suppose you shared them out between two children, or three children or
four children. When do you get some left over?

COUNTDOWN TO ZERO

Number patterns/ideas   Patterns   Counting

Count down from 10 to 0 or from 100 to 0.
Try counting in 5s, 10s or 20s.
When you get there, talk about zero. Is it nothing? What is nothing? Can
nothing be anything? What happens when you add 0 to a number? What
happens when you write it on the end of a number you have already
written down? What happens when you multiply by 0?

MAKING MONEY

Money   Addition   Subtraction

This game helps children to recognise coins and to practise addition
skills. You can make it easy or more difficult but make sure that children
succeed most of the time.
Have a selection of coins, 10p and less.

Make up simple problems:
    Can you make 3p with two coins?
    Can you make 8p with three coins?
    What is the fewest number of coins needed to make 9p?

A variation is to choose three or four coins and ask the children to take away one coin to leave a certain amount.

For example:
    Take away one coin to leave 11p or 6p or 15p.

With older children you could include 20p and 50p coins, and invent problems using five or six coins. There is little value, however, in playing this game with amounts greater than £1.

HOW MUCH?

Money   Language/talking   Sorting/classifying
Logical thinking

Suppose you are to be given twenty pence. What coins could be used to make the 20p?

One possibility is 10p + 5p + 2p + 2p + 1p.

Suppose you were given another twenty pence each time you discovered one more possibility.
How much could you receive altogether?

Begin by using real coins to provide a practical focus to this activity. But as strategies develop, try using paper and pencil or a calculator.

MEASURE YOURSELF

Length/distance

Look at the label in your coat. It will probably give a height in centimetres.
Are you the same height as on your coat label?
How tall are some of your favourite toys?

Shoe sizes are not given in centimetres.
Turn your shoe over and see if you can guess how many centimetres long it is from heel to toe.
Measure it to see if you were right.
Do you think your foot is the same length as your shoe?
Find out.

PLASTIC BOTTLES

Language/talking   Capacity/volume   Number patterns/ideas

Have you ever noticed the variety of plastic bottles in your home? Fill the kitchen sink or bath with water, provide a few bottles of different shapes and sizes and let your child play. Don't forget to try to challenge their thinking. For example:
    How can we find out which bottle holds the most?
    Which bottle holds the least?
    Can you estimate how many of the smaller bottles will fill that large one?
    Are there any bottles that hold the same amount but are different shapes?

Older children might like to compare the amount of liquid in bottles.
One bottle contains a litre of shampoo. What other bottles in your house contain a litre of something?
Make a collection of bottles and arrange them in order.
What is the least amount a bottle contains?
What is the largest amount?

JOURNEY BY BUS

Language/talking   Money   Time

Look at a timetable and decide what bus you need to catch.
What time is it now?

How long will it take to get to the bus-stop?
What time will we need to leave the house?
What is the number of the bus?
Is there a queue at the bus-stop?
Who is first in the queue?
Who is second, third, last?
How much is the bus ticket?
If you do not have the exact fare, do you know how much change you will get from 50p or £1?

TELLING THE TIME

Language/talking Time

Telling the time can be a difficult concept for young children to acquire. Breakfast-time, school-time, dinner-time and television-time are events in the day which mean something to young children.

'It's a quarter to nine – time to go to school.'

Look at the clock and point to the hour and minute hands as you say this. This will provide a familiar context in which children can learn to tell the time.
Using the clock-face and counting in ones, fives and tens forwards to and backwards from sixty, may also be helpful.
Ask your children to close their eyes while you say 'The big hand is pointing to six. What time is it?'

'It's half-past . . .'

With older children it is useful to compare analogue (the usual type of clock) and digital displays of time.

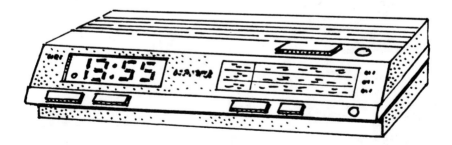

The digital display on the clock radio is 13:55. What time is it? In what position would the hands on a clock-face be?

HUMPTY DUMPTY

Shape/size

This is the famous Chinese Egg Puzzle.
Cut out the shapes carefully.

Move the shapes around and see how many different birds you can
make.

42   Try to put the egg 'together again'.

HIT FOR SIX!

Addition Subtraction Multiplication Division
Calculator use

You will need a calculator.

You may press only 4, +, −, ×, ÷, =

Use these keys only until you reach 6 on the display.
How many key presses did it take?

WRAPPING PAPER (track game)

Counting   Number patterns/ideas

You will need wrapping paper which has a repeated pattern or which has a scene or picture on it.

The wrapping paper with a scene, e.g. Postman Pat, can be pasted on to card and made to form a track for the game. The children can help choose where the game should begin and end and the stops along the way. The track can be made as long as you like and be marked in order with numbers so that the child may not only count forward and backward in ones but in even or odd numbers, threes, fives, etc.

WHAT'S THE DIFFERENCE?

Sorting/classifying

Choose two objects.

Ask the children to tell you all the differences they can find between the two objects. It could be differences in colour, shape, length, weight or texture, but they are sure to find all sorts of differences which adults would never think of!

TARGET

Addition Subtraction Multiplication Division

You can play this game anywhere, and adapt it for children of any age.
All you need is a paper and pencil.

Choose three or four numbers and start with just addition and
subtraction. For example:

2, 3, 5, +, −

Now give a 'target' number and try to reach it, e.g.

10   3 + 5 + 2 = 10
6   3 + 5 − 2 = 6
4   5 + 3 − 2 − 2 = 4

For older children allow multiplication and division as well. For example:

Target number is 30   3 × 5 × 2 = 30
Target number is 13   5 × 3 − 2 = 13
Target number is 70   35 × 2 = 70

You can change the numbers and vary the rules. For example:

only use each number once
only use each sign once (+ − × ÷)
you must use every number

In the home

7- to 11-year-olds

SQUARE PUZZLE

Logical thinking

Set out 24 sticks to make squares as shown below.

Try the following:
1 Take away 8 sticks and leave 2 squares.
2 Take away 6 sticks and leave 3 squares.
3 Take away 4 sticks and leave 5 squares.
4 Take away 8 sticks and leave 5 squares.
5 Take away 8 sticks and leave 3 squares.
6 Rearrange all the sticks to make 1 large square.
7 Rearrange all the sticks to make 2 squares.
8 Rearrange all the sticks to make 3 squares.

CONNECT FOUR (a game for 2–4 players)

Multiplication

You will need a set of dominoes with the blanks removed, and a piece of paper marked with a 6 by 6 square numbered 1 to 36.

1	2	3	4	5	6
7	8	9	10	11	12
13	14	15	16	17	18
19	20	21	22	23	24
25	26	27	28	29	30
31	32	33	34	35	36

Place the dominoes face down on a table. Players take it in turns to chose one. Multiply the score on the two halves together and cross off the resulting number on the marked square. For example, a double 4 domino means 16 can be crossed off, a 2-dot and 5-dot domino means 10 can be crossed off.

The first player to connect 4 numbers in a line, vertically, horizontally or diagonally, wins.

Variations and extensions

1 Number the square board differently.
2 Use a rectangular board 9 by 4.
3 Try to connect 4 numbers to form as many squares as possible on the square board.
4 Use the blank dominoes.

TOTAL 100

Addition   Investigating

Try to use each of the digits from 1 to 7, using each once only, to make an addition sum that has a total of 100.

SWEETS

Division   Investigating

When a girl counted her sweets into piles of four, she had two left over.
When she counted them into piles of five she had one left over.
How many sweets did she have? 2 1
Are any other answers possible? 7 in 3 plus 3 in 7 plus

Can you make up other problems of this type?

HOLIDAY PACKING

Arranging

Ever had arguments about how much to pack?
The children want to wear something different every day. The question is how much is needed?

Put out two t-shirts and two pairs of shorts and ask:
   How many outfits? Is it two or more?

Put out another t-shirt or pair of shorts and ask:
   How many outfits now?

If there were four t-shirts and three pairs of shorts how many different outfits could be worn?
Is there enough for a fortnight's holiday yet? If not, how many t-shirts and shorts would be needed for each child?

EXPLORING NUMBERS

Language/talking   Estimating/approximating   Area

Talk about very large numbers and very small numbers.

Cut a piece of paper in half, and again and again . . .
Could you carry on for ever if you had special equipment?

How could we count the stars in the sky or the grains of sand on a beach
or the blades of grass on a lawn?
Can we work out a sensible estimate?

Try looking at a photograph of a large group of people.
Do we need to count every one?
Suppose we count the faces in just one part of the picture.
How many similar parts are there?
Does this help you to estimate the number of people in the whole
picture?

USING A CALENDAR

Time   Number patterns/ideas   Counting   Mental calculation

Encourage the children to use a calendar regularly so that they know the
dates and days of family birthdays and holidays. Ask them to say whose
birthday is next and how long it is until that date.

Do the same for other important events that the children may be looking
forward to: Christmas, Easter, holidays or visits. Look for interesting
dates:
13 May 1988 is a Friday. Is there another 'Friday 13th' in that year?
On what day does 1 April fall this year?
Is it on a weekend, during the week, holiday time or schooltime?

Encourage mental calculation with questions like:
Today's date is 4 July.
What day of the week is it?
What date will it be on the same day next week, or last week?
What will be the date next Saturday?
What was the date the day before yesterday?
What will the date be a week on Wednesday?
How many Tuesdays this month?
How long till the end of term?
What day do you go back to school?
How long is the summer holiday?
We go away for the first two weeks in August and Gran would like to
visit for a week while you're home from school. Is there time before we
go away?

MAKE A CONTAINER

Estimating/approximating   Shape/size   Capacity/volume

You will need a collection of empty packets and boxes, scissors, a pencil, ruler, paper, felt-tip pens or colouring pencils, and glue.

Pieces of a puzzle or game are often lost because of a broken box. Make a container into which the game or puzzle can fit and be removed easily. Use the empty boxes to give you an idea of a good design and shape. Cut them and open them out flat so that you can see how they were made and assembled.

When the container is made cover it with paper. Either stick the old label to it or design one of your own.

QUICK ON THE DRAW

Addition   Subtraction   Multiplication   Division

This family game encourages the quick recall of number facts. It can be adapted to suit children of any age and can incorporate any number facts or tables that need practice.

At its simplest level – number bonds to 10.
Two members of the family stand back to back. The rest of the family and friends take it in turns to call out a number less than 10, e.g. 7. The first one to give the number that brings the total to 10, i.e. 3, 'shoots down' the other player and lives to fight another time. That player remains standing and another member of the family comes to stand back to back with them.

Variations
1 Use larger number bonds – to 20, to 50, to 100.
2 Practise tables by making it a '3 ×' game. Whatever number (up to 10) is called out, the 'gunfighters' must multiply it by 3.
3 As in the previous game but divide by a chosen number, e.g. 4. The number 24 is called out – first one to shout 6 wins.
4 A 'take 5' game. Whatever number is called out, the gunfighters must subtract 5. E.g. 13 is called out – first to shout 8 wins.

CROSSING THE RIVER

Investigating   Logical thinking

Two men and two boys want to cross a river.
None of them can swim and they only have one canoe.
They can all paddle but the canoe will only hold one man and two boys.
How do they all get across?

PASSWORD

Addition   Subtraction   Multiplication   Division

Every time your child (or anyone) goes out of the room, or upstairs, or asks for something, they must give the right answer or password. For example, they are asked to calculate 7×6 and the 'password' 42 must be given.
Your child could do the asking of everyone else – checking their replies.

It is better to concentrate on one number fact for, say, a day (and come back to it a week later) rather than try to learn several at the same time.
Remember to use its reverse, e.g. 7×6 and 6×7.
You can also bring in division, e.g. how many 6s in 42.

WHICH WAY UP?

Probability/statistics   Predicting

When a drawing pin is dropped on to a flat surface, it either falls with its point up or down.

Which is more likely? How many times must you try before you can be reasonably sure of being right?
If a box of 100 drawing pins fell off the desk, how many would land point up?
What else can you find to investigate in this way?

51

THEY SAY ... (1)

Length/distance   Ratio   Average

They say that your height is about three times the distance round your head.

Is this true ... for children?
            for adults?
            for babies?

Card Games

For all the following games use an ordinary pack of cards with Jacks, Queens and Kings removed – giving a pack of 40. Aces count as one. In most games it does not matter if one or two cards are missing.

ADDSNAP

Addition

The dealer shows two cards.
They are won by the first player to call their sum (+) correctly.
This is repeated until all the cards are won.
The player who gets the most cards is the winner and becomes the dealer for the next game.

TIMESNAP

Multiplication

This is played like *Addsnap* but players have to call the product (×) of the two cards.

ELEVENSES

Addition

This is a game for one player.
Deal eight cards face up and look for pairs of cards which total eleven. When you find a pair, deal a new card on to each card in the pair. Continue finding 'pairs of eleven' and dealing new cards on top until either you have used up all the cards – in which case you win, or until you cannot go – in which case you lose.

LAYOUT

Multiplication

The 40 cards are dealt face up to make a grid 8 × 5.
The dealer chooses two cards that are next to each other, either
horizontally or vertically, and, without saying which they are, calls out
their product.
The other players try to find them (or two other cards that give the same
product).
The first player to point to a correct pair of cards wins those two cards.
This player chooses another pair of cards and calls out their product.
Play goes on like this until all the cards have been won.
The winner is the player with most cards.
(Sometimes remaining cards will need to be picked up and re-dealt.)

SHOW MOST

Place-value

Deal ten cards to each player. (In this game tens count as 0.)

Round 1
You all choose a card from your hand and put it face down on the table.
When everyone has done this, the cards are turned over. The player
showing the highest value wins 1 point. If two or more players have the
same highest value no points are given.

Round 2
Now you all put two cards face down, arranged side by side to show the
highest value (e.g. 2 and 7 would be arranged to show 72). When the
cards are shown the player with the highest value gets 2 points.

Round 3
Three cards and 3 points.

Round 4
Four cards and 4 points.

At the end of the hand the cards are collected and re-dealt.
The first player to get 25 points is the winner.

Variation
Call the game 'Show least' and aim to make low numbers.

SEQUENCES

Patterns   Predicting

Pattern is often a clue to the solution to a problem (not just in mathematics). Once you have spotted a pattern you are then able to predict what will happen next.
Try this sequence:

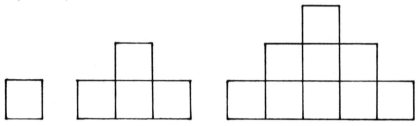

Can you see how the shapes are growing?
Can you predict the number of squares in the next shape up?
Can you draw it?
There are many ways to work on this problem. You can draw, use plastic cubes if you have them, or look for number patterns. There may be other ways.

How about this one?

 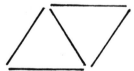 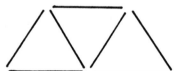

1 triangle takes 3 sticks

2 triangles take 5 sticks

3 triangles take 7 sticks

Can you say how many sticks you would need for 10 triangles?
How about 100 triangles?

GUESS WHAT THE COMPUTER IS DOING

Addition   Subtraction   Multiplication   Division   Patterns

Any number of people can play this game.

One of you is the computer. You decide which mathematical operation(s) you will perform, e.g. +1, +10, −3, ×2+1, etc. Don't tell the others which operation you have chosen. Other players feed you numbers. You perform your chosen operation and deliver the answer.

Example:
You have decided to add eight (+8).
Players in turn say 2, 4, 5, 8. You respond in turn with 10, 12, 13, 16.

The person who guesses the rule (i.e. your mathematical operation) becomes the computer.

REFLECTIONS

Symmetry

Find pictures in magazines, catalogues, books, etc. of familiar objects that have either vertical or horizontal line symmetry. (A line of symmetry is when a line can be drawn which divides the picture into two identical halves, each a mirror image of the other.)

Cut the picture in half through the line of symmetry and stick one half onto a piece of paper.
Ask the children to draw in the missing half to complete the picture and then use a mirror to see how accurate they were.

Letters of the alphabet can be sorted into those that are symmetrical and those that are non-symmetrical. Make up mirror words like:

MAKE 15

Addition   Logical thinking

The object of this game is to make a row or a column or a diagonal total 15 by placing number cards onto a grid.

The grid used is:

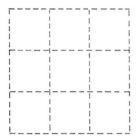

The numbers 1–9 are used only once.

Player A has cards with the numbers 1, 3, 5, 7, 9.
Player B has cards with the numbers 2, 4, 6, 8.
A plays first, then B, then A, and so on.
The player who makes 15 first is the winner.

LARGEST PRODUCT

Multiplication   Estimating/approximating   Investigating

Use the digits 1–5 to make a multiplication sum.

What is the largest answer you can make?
What is the smallest?

Try the same problem using the digits 5–9.

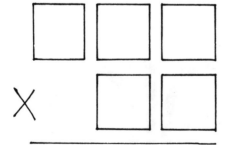

REFLECT ON THIS

Symmetry

Can you place a mirror on AB to find (a), (b), (c), (d), (e) and (f)?
When you have found them all, use your own initials to make similar
puzzles. Try them on each other.

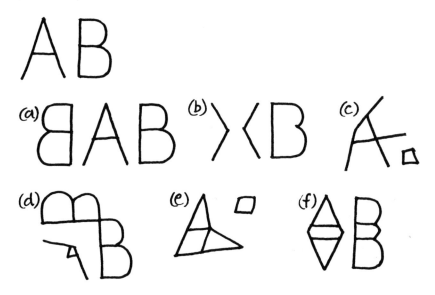

CONSECUTIVE NUMBERS

Number patterns/ideas   Logical thinking

2 is consecutive to both 1 and 3.
3 is consecutive to both 2 and 4.

In the diagram below, arrange the numbers 1–5 so that no two
consecutive numbers are joined by a line.

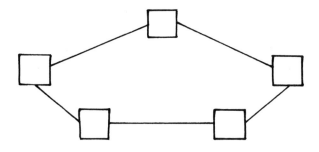

RICH FOR A DAY!

Number patterns/ideas Money Addition Subtraction
Multiplication Estimating/approximating Calculator use

This game can be played by any number of players. You will need plenty of newspapers and mail order catalogues, pencils, paper and calculators.

The rules of the game are:
   No money can be given away.
   You cannot buy more than two of the same item (e.g. cars or houses).
   You are allowed to round prices up or down to the nearest £1 or £10.

Everyone starts with £1000. They 'buy' items from the catalogues or newspapers and keep a running subtraction of purchases until all their money is spent.
If more than one person spends their £1000 at the same time then the person who has bought the most items is the winner.

Variations

1 Increase the amount to £10 000 or £1 000 000 – a surprise pools win.
2 Include a telephone directory with the cost of international calls. (Restrict players to no more than two phone calls of less then ten minutes each.)

WHAT DAY DID IT HAPPEN?

Addition Division: with remainders Number patterns/ideas

Choose a date, e.g. 25 December 1945	
Use the last two digits of the year	45
Divide by four (ignore any remainder)	11
Add these two numbers together	56
Add the date	25
	81
Add the number for the month, December (see *Table 1*)	6
	87
Add the number for the year, 1945, (see *Table 2*)	0
	87
Divide this by 7	12 remainder 3

The remainder gives the day of the week (see *Table* 3).

Table 1
January – 1 (0 for a leap year);
February – 4 (3 for a leap year);
March – 4;
April – 0;
May – 2;
June – 5;
July – 0;
August – 3;
September – 6;
October – 1;
November – 4;
December – 6.

Table 2
For any years beginning 18 . . – add 2;
for any years beginning 19 . . – add 0;
for any years beginning 20 . . – add 6.

Table 3
Sunday – 1;
Monday – 2;
Tuesday – 3;
Wednesday – 4;
Thursday – 5;
Friday – 6;
Saturday – 0.

So, 25 December 1945 was on a Tuesday.

Try finding the day members of the family were born and see if this calculation always works.
Look ahead and calculate the day Christmas or any other celebration will fall in the year 2000. (Remember to work out if it's a leap year.)

HOLIDAY CHOICE

Money   Logical thinking

Provide a selection of holiday brochures and set children imaginary (or real) challenges to find the right holiday. For example:
'I want a self-catering holiday abroad for a family of four. I would like to fly from the local airport but I don't want to be travelling for more than five hours. I want to go in the first week of May and the apartment must have use of a private swimming pool and be near the sea. I have £600 to spend. Where can I go?'

Children could choose their own favourite holiday – and tell you why they chose it.

You can give a similar challenge for holidays in this country. Include the amount of money to spend on petrol and the m.p.g. figures for your car.

Is spending money included in these estimates?

LETTER PUZZLES

Addition   Logical thinking

For the letter M, use the numbers 1–9.
For the letters A and S use 1–6.
For the letter T use 1–5.
For the letter H use 1–7.

For each letter puzzle, place one number in each space. The aim is to make each line add up to the same total.

You may find it quicker to copy out each puzzle on to a large piece of paper and write the numbers on small pieces of card or paper. You can then move the numbers about.

How many different ways can you do each puzzle?

One puzzle can't be done.
When you've decided which one this is, try to say why it can't be done.
Try this with other letters.

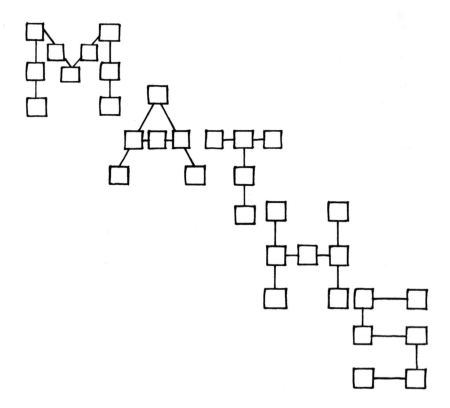

FIRST TO 20

Addition   Logical thinking

Two players share a pencil and paper.
Play starts at 0.
Each player adds on, in turn, any number between one and five.
The winner is the player who is able to 'land on' 20.

Extend this up to 100 or down from 100 and bigger jumps using a calculator.

PLOTS

Logical thinking   Shape/size   Area   Multiplication   Division

A builder has bought these three square fields.

He wants to build eight houses. How can he divide up his fields so that all the plots are the same size and shape?
If the side of each field is 100m, what will be the area of each plot?
Can you devise a puzzle like this of your own? Try it out on your friends.

THEY SAY ... (2)

Logical thinking

They say that a map showing the countries of the world can be coloured in such a way that, using only four colours, no two countries of the same colour will ever be neighbours.

Is this true?
Draw your own map and try it.
Note Use coloured counters first before you colour in. Counters can be moved about!

SQUARE CORNERS

Angle/direction   Estimating/approximating

You need any piece of scrap paper, e.g. an old envelope or newspaper.

To make your square corner:
1 Fold the paper roughly in half.
2 Crease the fold firmly.
3 Fold this fold exactly along itself.
4 Crease this fold firmly.

Keeping the paper closed, the square corner (right angle) should be clearly seen against the two creased folds.

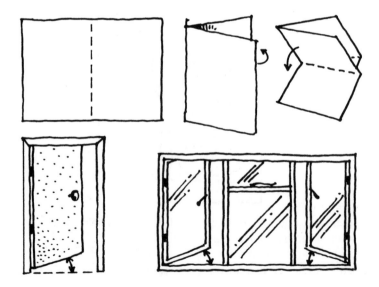

Make a list of where you think you might find right angles in the home. Which sorts of place would be good to begin with?

Check with the square corner and see if your list was correct. What can be added to it?

A folded sheet of newspaper can have its fold firmly creased. It can be folded along this crease to make a large right angle.

Children can then be asked:

Which square corner is best to check floors and doors?

Which to check books and packets?

Can you find right angles which are upright and some which are on a flat surface?

Try to list articles which have angles larger or smaller than right angles. Some types of packaging are good for this.

HIDDEN SHAPES

Shape/size   Patterns   Investigating

There are five squares in this figure.

Can you find them?

How many in this?

And this?

How many squares are there on a chessboard? (A chessboard is 8 × 8.)
Can you find a way of working this out without counting?

Now try with triangles

  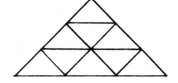

DEAD ON TARGET

Addition Subtraction Multiplication Division Place-value Fractions Calculator use

Any number of people can play this game. You will need a pack of playing cards with the picture cards removed, a pencil and paper, and a calculator.

Shuffle the cards and place the first two face up on the table. These form the target number, the first card having its value in tens. So, if the 3 of diamonds is turned over, followed by the 9 of hearts, the target number is 39. If the first card turned over is a 10, it has the value of 100; if the second card turned is 10, it has the value of 10.

Deal the players five cards each. They use these and the four mathematical signs ($\times$, $+$, $-$ and $\div$) to make the target number.

Target number 110

Target number 113

Target number 13332
King = 1300
3 clubs = 30
2 spades = 2

Example:
If I am dealt a 5, 4, 10, 7 and 7 as my five cards, and the target is 39, I could put down:

$5 + 4 + 10 + 7 + 7 = 33$

or

$(10 \times 7) - (4 \times 7) - 5 = 37$, which is closer to 39.

The aim is to get 'dead on target', or as close as possible.
Score as follows:

'Dead on target' – score the target amount.
Within five of target – score half the amount.
Within eight of target – score a quarter of the target amount.

Variations

1 Keep in the picture cards; Jacks = 11, Queens = 12, Kings = 13.
2 Turn three cards to give a three-digit target number. If the first card is 10, the target number will begin with 1000.
3 Players can ask for a sixth card but the value of this additional card must be subtracted from their final score.

HIT THE TARGET

Addition   Subtraction   Multiplication   Division

You will need a calculator.

You are given two digits (say 3 and 8), the +, − and = signs. You may press them in any order and as frequently as you like to hit a given target in the least number of key presses. Ask someone not playing to choose the target between 10 and 20 (say 15) or write several targets on pieces of paper and draw one at a time from a bag or box.

Example:
$8 + 8 + 8 - 3 - 3 - 3 = 15$ (score of 12 key presses).
   or $3 + 3 + 3 + 3 + 3 = 15$ (score of 10 key presses).
Remember   Lowest score wins.

Variations and extensions

1 Use larger/smaller targets.
2 Use +, − and ×.
3 Use all four operations.
4 Use more number keys.
Can you produce all the numbers from 1–9 or 1–20?

FAULTY KEYS

Addition Subtraction Multiplication Division Calculator use

Some of the keys on a calculator are stuck.
The only number keys that are working are 5 and 7 but all the function keys (+, −, × and ÷) are operating normally.
Using only the keys that are working, try to make all the numbers from 1–20.

Example:
12 could be made by 7 + 5 = 12.

POCKET-MONEY

Addition Multiplication Ratio Estimating/approximating Predicting

Let's think about pocket-money.
Two children tried to persuade their parents to start giving them pocket-money.
They said: 'We're not asking for a lot. We'll start with 1p a week.' (They thought their parents would like that!) 'But our pocket-money must double every week.' (Will their parents like that?)
Work out what happened in week 1, week 2, week 3, and so on.
Keep a table of your results.
When did you first need to use your calculator?
In which week do you think their parents said 'STOP!'?
What do you think would happen in your family if you tried this?

PERFECT NUMBERS

Addition Division Investigating

6, 3, 2 and 1 are *factors* of 6. They go into 6 exactly.
3 + 2 + 1 = 6.
Six is called a *perfect number* because it is equal to the sum of its factors.

Can you find any other perfect numbers?

TANGRAM

Shape/size   Area   Investigating   Problem-solving

This is an ancient Chinese activity.

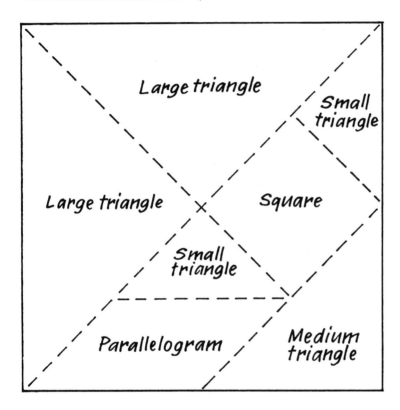

Start with a square of card and cut out 7 pieces, as shown. Try the following:

1 Put all 7 pieces together to make the square again.
2 Put all 7 pieces together to make a rectangle which is not a square.
3 Cover the small square with 2 small triangles.
4 Cover the medium-sized triangle with 2 small triangles.
5 Cover the parallelogram with 2 small triangles.
6 Cover a large triangle with 2 small triangles and the square.
7 Cover a large triangle with the parallelogram and 2 small triangles.
8 Put all 7 pieces together to make 1 large triangle.
9 Cover a large triangle with the medium triangle and 2 small triangles.
10 How many different designs can you make with the shapes? Try a boat or a face to start with.

WEATHER COUNTDOWN

Collecting/using information   Temperature

Two weeks before a trip, keep a record of the daily temperature at home. Encourage this to be checked against weather reports on the TV or in a daily newspaper. At the same time keep a record of the daily temperature reported for your proposed destination.

If there is no report of the temperature at your destination, look for the temperature of the nearest resort, or for the region in which your destination is situated.

Ask questions like:
Where is it warmer?
How much warmer is it?
Is the temperature at home remaining much the same or changing?
If it's changing, is it getting colder or hotter?
Do you think it will be warmer at home or away for the time of the trip?
How close is the actual temperature at home to the one given by weather forecasters?

If possible listen for, or ring the weather bureau for, the long-range weather forecast.
Keep a note of what the forecast says and, if possible, keep a temperature check whilst you are away.

DO-IT-YOURSELF

Estimating/approximating   Length/distance   Capacity/volume Area   Money

Next time there's a room to decorate, let the children help with the planning.
They can measure the size of the room and, with a calculator, estimate how much wallpaper will be needed. They will need to look at the pattern to see whether it needs matching, and allow for this.
They can use similar skills to estimate the amount of paint needed, the number of tiles for a kitchen or bathroom, and the length of carpet or floor covering.

If you set a price limit, children can budget to stay within the amount specified.

HOW FAR WILL IT OPEN?

Angle/direction

For convenience most doors open against a wall. They therefore only open to 90° or a right angle.

Is this true of your house? Check interior and exterior doors, especially those that do not open against a wall. Think about the doors of cupboards, trap-doors into a loft or desk lids which may open upwards.

Record all those which open to form a right angle (90°).
Do any open as far as two right angles (180°)?
Record any that open between 90° and 180° (obtuse angles).
Windows are a good source to study for angles. (Keep to those that are easily accessible downstairs.) Will any open as far as 180°?

Many windows have adjustable bars which will hold them in place once opened.

Make a table showing how many positions the bar has which allow the window to remain open at an angle less than 90° (acute angle), or between 90° and 180° (obtuse angle).

SLEEPY HEADS

Time Addition Subtraction Multiplication Fractions Decimals/percentages

As a family, you can work out how long you spend asleep. Each member of the family should give the time they usually go to bed and the time they usually get up.

Use a calculator to work out the total time each person sleeps and then add these together to calculate the total time the family sleeps.
How long does your family sleep in a week?
How long in a year?
Make a chart to show the different number of hours each member of the family spends asleep.
What fraction of the day do you spend asleep?
What percentage of your life will you spend in bed?
If you turn in your sleep once every half hour, and you go to bed at 21.30 and wake up at 7.30, how many times will you turn over?

BODY MATHS

Estimating/approximating Shape/size
Collecting/using information Length/distance

You will need a tape measure or ruler, pencil, paper and a co-operative family.

Measure as many members of the family who are willing, including yourself. In centimetres, measure:
The distance around the head.
The height.
The distance from fingertip to fingertip with arms outstretched.

Once this information is recorded, find the answers to the following:
Is the distance around the head three times that of the height?
Is the height approximately the same as the stretch?

Before answering these questions ask:
Is it likely that anyone's corresponding measurements will be exact?
What approximations would be acceptable – within 3cm, 5cm or 10cm?
Whose measurements are most likely to correspond – the oldest members of the family or the youngest?

TV ADDICTS

Time   Addition   Subtraction   Multiplication   Calculator use

Make a survey of which member of the family watches the most
television. Make a list or a chart for each member of the family for a
week.

Before you go on holiday, choose the programmes you would like to
video while you are away.
Can the children set the clock?
Ask them to estimate what length of tape will be the most economical.

THUNDERCLAP

Time Division

How far away is the storm?
Watch for the lightning and count the seconds until the thunder is heard.
Put the seconds into groups of three. Every 3 seconds indicates that the
storm is 1 kilometre away. (Sound travels at 330m/sec.) If you count 13
seconds the storm is about 4 kilometres away.

Outdoors

4- to 8-year-olds

WALKING TO SCHOOL

Language/talking Estimating/approximating Length/distance

Could you give a stranger clear directions of how to get to your school?
To do this you need to be aware of noticeable landmarks, e.g. shops,
zebra crossings, traffic lights, etc.
Do you have to cross any roads or turn left or right?
Perhaps you could make a pictorial map to help you.

ROBOTIC WALKS

Shape/size Angle/direction

Find a safe place in your garden or a park. Blindfold a friend and then
give simple instructions to direct the person to a tree or bench. Here are
some commands to help you.

Example:
Forward   6
Right
Forward   4
Left
Forward   3

What instructions would you need to give to face your friend in the
opposite direction?

STEPPING OUT

Language/talking Estimating/approximating Length/distance

An activity such as this is designed to promote an understanding of
length and distance. The important skill of estimating improves with
practice.

Let's count how many steps it takes you from this lamp-post to that one.
Can you guess how many steps to the next lamp-post?
I wonder if the number of steps is the same between all the lamp-posts?
Now let's count how many steps it takes from . . . to . . . , etc.

SIMON SAYS

Shape/size Language/talking

A game that can be played anywhere. A number of shapes can be marked on the ground or beach, for any number of players.

Draw out some large shapes, such as circles, triangles, squares, pentagons, etc. They need to be large enough for as many players as you have to stand in them.

Now play 'Simon says', giving instructions to use the shapes.
Example:
Simon says jump in the square,
        hop in the triangle,
        walk around the hexagon,
        sit in the circle.

Some instructions should not use the name of the shape. This will require the players to choose as there may be shapes with identical features.

Example:
Simon says sit in a shape with four sides,
        stand in a three cornered shape,
        walk around three square corners.

AT THE GREENGROCER'S

Language/talking Shape/size Number patterns/ideas
Patterns Weight Money Sorting/classifying Counting

A visit to the greengrocer's provides many opportunities to 'talk mathematics', not only at the shop, but at home also.

At the greengrocer's
Which things belong to the family of vegetables?
What is the biggest fruit you can see?
How many oranges can we buy for 50p?
Does anything we buy need to be weighed?
Ask the greengrocer to show you the weighing scales. See some things being weighed, and watch the dial move. Where is the mark for 1 lb, 2 lb, 1 kilogram?

At home
Find out which is the heaviest potato; the lightest potato! Are any the same size?
Can you find two things that are the same size but not the same weight?
Can you find two different types of vegetable or fruit that are both about the same size and weight?
Many fruits and vegetables have interesting patterns when you cut them in half.
Look at onions, cabbages, oranges and apples.
Count the segments in an orange.
Do apples always have the same number of pips?

SPOT THE SHAPE

Shape/size

How many shapes can you spot on buildings, walls, pavements, etc.?
Apart from the obvious squares and rectangles, once you start looking
you often see:

    circular and semi-circular windows
    hexagonal patio paving slabs
    triangular roof gables

You can also spot 3-dimensional (solid) shapes:

    cylinders – chimney pots, pillar boxes
    cones – traffic cones (are they complete cones?)
    spheres – stone spheres on top of gateposts
    prisms – some tents
    pyramids – roofs, some tents, some church spires

You could adapt this activity to an 'I spy' game:
'I spy, with my little eye, a shape beginning with . . .'

PATTERN PUZZLES

Patterns   Logical thinking

Ask the children to take a wax crayon (or a soft pencil) and some paper
with them on a trip. During the day they make rubbings of various objects
they see, for example, tree trunks, gratings, walls, the pavement, different
surfaces, leaves, inscriptions on walls.

Back at home, can they guess what their friend's or their brother's or
sister's rubbings are from?
What are the clues to where the rubbing came from?
Which materials make the best rubbings, hard things or soft things?

WASH DAY

Sorting/classifying   Patterns

The weekly wash can be used for sorting activities with young children.

Sort the washing into pale and strong colours ready for the washing machine. Can they use the 'care labels' to help sort the clothes into hot wash or cool wash?

When you hang out the washing, invent a game where they pass you the items according to your rules:

> The same type of clothes cannot be hung next to each other.
> The same coloured clothes cannot be next to each other.
> Make a repeating pattern – plain clothes, patterned clothes.
> Repeat a pattern of clothes, e.g. those worn above the waist, below the waist, on the feet.
> Choose a simple pattern yourself, but don't say what it is.

Can the children repeat your pattern? Listen to their explanation because, although it may not be the pattern you thought of, it could still be right!

LOST AND FOUND

Angle/direction

The children close their eyes while you hide a small object in the grass or sand. You then direct them to the object by calling out instructions, for example, forward 3 paces, turn left, forward 6 paces, turn right, back 2 paces, etc.
Change places and let the children hide something and direct you to find it.

For older children, you can make the route more complicated. Use the terms north, south, east and west and include the number of degrees in each turn, for example, right 90°, left 45°.

ODD-ONE-OUT

Number patterns/ideas   Shape/size   Length/distance   Time

You can play this game anywhere – in the park, on a long journey, at the seaside.

Give three or four numbers and ask for the odd one out.

Example:

8	4	7	2	it could be 7, because it's an odd number
6	12	9	4	it could be 9, because it's odd, or 4, because it's not in the 3× table, or 12, because it's more than 10.

You could play the same game with shapes. For example, point to three rectangular windows and one square one.
It could be the odd one out in colour, size, length, time, etc.

Another idea is to make up a simple number sequence and deliberately include one that doesn't fit.

Example:1   3   5   7   9   *10*   11   13   15   17

BIRD WATCHING

Sorting/classifying   Patterns   Probability/statistics

Watch the birds that visit your garden or neighbourhood. You may need a bird book to help you identify them.

Study the food they eat and where they build their nests. Set aside a time during the day to make a survey. Record your information by keeping count of the number of visits of each type of bird.
Which birds come most often?
Which birds rarely come?
Talk with your child about these things.

THE 'GUESSTIMATION' GAME

Estimating/approximating Length/distance Time Weight

This game is fun and very good practice in the important mathematical skill of estimation. What you choose to estimate will depend on the stage the children have reached. Remember, it will only be fun if the children can succeed most of the time. Here are some suggestions:

Say 'GO' and the children shout 'STOP' when they think a minute has passed. Tell them if it was too long or too short and try again another day.

Guess how far, in metres, to an object (not too far away to begin with) like a post-box or a lamp-post. A child's stride will be about half a metre.

Estimate the height of a door, the width of a window, the length of the garden. (A door is about 2m high.)

Estimate the weight, in kilograms, of members of the family and friends. (A man of average build, height 1.8m (6ft) weighs about 76kg (12 stone).)

In the car, estimate how far you have travelled (use the milometer reading to check). What speed are you doing? How far is it to . . . ?

BUILDING A SANDCASTLE

Estimating/approximating Logical thinking Counting

You will need a collection of buckets and spades and/or plastic pots, e.g. yoghurt, margarine or ice-cream tubs. Shells.

Ask the children to make a castle to your specifications.

Example:
The castle must be a certain shape: triangle, circle, square or rectangle.
Each side must be as long as something: three spades, deckchair, a towel.
It must have four large turrets. On top of these, two smaller turrets must be placed.
Between these tall turrets, along each side of the castle, there must be room for three of the smallest sand pies.
Each turret should have six windows (shells) all the same size! Only the correct amount of shells should be used.
Other additions could be the depth of the moat, the number and height of doorways, etc.

FLOWERPOTS

Estimating/approximating Arranging Patterns Symmetry

You will need a collection of clean, empty flowerpots.

Turn two upside down. Ask:
Can one be balanced on top?

Place three pots side by side. Ask:
How many can be balanced on top now?

Place the three pots to form a triangle. Ask:
Can the same number be balanced on top now, or more, or less?

What happens if three large flowerpots are chosen and smaller ones balanced on top?
What happens if four pots are chosen?
Try them in a line, or in a square at the base.
Try some about the same size and those that are larger at the bottom.

What's the largest number that can be balanced on four pots?
Are there enough to place six at the base?
How many ways can you arrange the six base flowerpots?
Which arrangement gives you the tallest stack?

A RAINY DAY

Language/talking Shape/size Capacity/volume

You will need wellington boots!
Watch raindrops fall into puddles. What do you notice?

Estimate how many footsteps it will take you to walk round a large puddle.
Find out if your estimate was correct.

Use a stick to measure how deep a puddle is.
Find a way of marking the stick so you can use it to measure the depth of other puddles.

Find an eggcup or beaker and collect water on rainy days.
Did you collect more water today than yesterday?

A SUNNY DAY

Length/distance   Angle/direction   Time

Stand outside your house and look at shadows. Where is your shadow? Where is the sun? Does the shadow of a tree have the same shape as a tree?

How could you make a tall shadow? A short shadow?
Where is the sun at midday? Where is your shadow? Where will the sun and your shadow be at 4 o'clock? (Such discussion could lead to finding out where north, south, east and west are.) In what direction does the front of your house face?
Look at a sundial and discuss where the shadow falls and what time it is. Can you make a simple sundial using a stick and stones?

PUB SIGN MATHS

Addition   Mental calculation   Subtraction   Place-value

This game is for passing the time on journeys that are not on motorways. You will need pencil and paper.

One child begins to spot pub signs. For each limb there is a point to be scored, e.g. The Fox and Goose scores six, The Green Man four, The King's Head or The Ship none.

If the pub spotted scores none, another member of the family has their turn to spot.

A target number can be set, say 50, or less depending on the interest in the game. The aim is to reach the target before arriving at the destination.

Alternatively, the pub scores could be subtracted from a figure. The aim of this game is to score zero.

82

WHEELS

Language/talking Shape/size

The wheel is considered one of our greatest discoveries. Nearly every machine we know has a wheel somewhere in its design.
Look around you. What are wheels used for?
Make a point on a bike wheel and roll it over the ground. What shaped path does it follow?

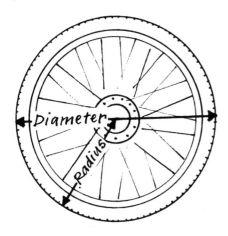

A cog is a wheel. Cogs are found in car engines, watches, etc. Find out how cogs work. Try to make a gear-system using corrugated cardboard and pencils and a pair of compasses.

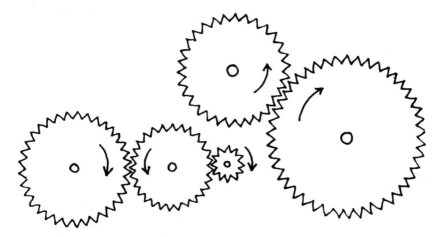

SPOT THE SIGN SHAPE

Addition   Collecting/using information   Shape/size
Multiplication

You will need a check sheet of the four signs, with a value given to each one.

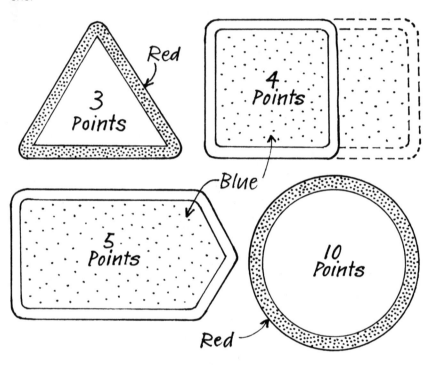

Give each child one side of the road to observe.

Ask them to make a tally of the signs they see during the journey, or part of it.

At the end of the game the children can use a calculator to work out their total number of points for each sign and then the final total for all signs. They can do this by using either addition or multiplication, or both.

The child with the highest score is the winner.

Variations

1 Collect the largest number of a particular shaped sign, e.g. circles.
2 Ask each child to look out for a different shaped sign. At the end compare their totals and decide whether circles, triangles, rectangles or pentagons are most frequently spotted.

HOPSCOTCH MATHS

Addition   Multiplication

These can be played anywhere a hopscotch pitch can be marked out.

Addition hopscotch

Each player takes it in turns to throw two stones or shells on to two numbers which make ten, beginning with one and nine, two and eight, and so on. The player hops along the pitch without touching the squares with the stones in, collecting the stones on the return. The first player to complete all the sums is the winner.

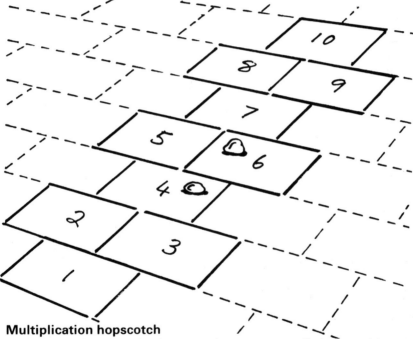

Multiplication hopscotch

The game is played as above, except the stones are rolled on to table facts in order. For example, if the tables of two are chosen, the stones are rolled on to two and one first. The number two is called by the player and the pitch is hopped, missing out squares one and two. The stones are picked up by the player on the return hop.

The player repeats this, working through the chosen table in order. If the player calls out the wrong table fact, or misses the correct square, the next player takes a turn.

The aim is to hop the whole pitch through one whole table to ten.

PLAYING IN THE PARK

Time Counting

Choose any piece of apparatus and try some timing activities. Here are some suggestions.

On the swings and roundabout ask:
    If I give you one push will you stop before one minute?
    Can you count the number of swings or revolutions caused by one push?
    Will two pushes give you twice as many?

Assuming there is not a queue at the slide, ask:
    How many slides can you do in one minute?
    Can you do twice as many in two minutes?

Choose something that the children enjoy, but are unsure of, e.g. the climbing frame. Time them climbing up one side and down the other. Let them have turns without timing and then, on another occasion, time them again to see if they can do it quicker. The aim is to make improving a skill fun and safe, so only time when the children are not tired or over-excited.

When time is short try asking:
    We only have five minutes – can you have a go on everything?
    We have ten minutes – choose two things and spend five minutes on each.

NUMBER SEARCH

Number patterns/ideas

Noticing numbers can be a source of fascination and enjoyment for young children. To some, one hundred seems a very large number but others may be ready to explore thousands and even millions. Naming and interpreting the numbers we see in everyday life can help children gain confidence in handling numbers. A few suggestions are outlined below but the possibilities are endless.

1 On one side of the street the houses are numbered 3, 5, 7.
    What will be the numbers on the next three houses?
    What is the number of the house between 110 and 114?
2 Look at some road signs that show the distance between two places.
    Can you work out how far it is to your nearest town or city?

3 Look at the prices of cars at a local garage.
  Which is the cheapest?
  Which is the most expensive?
4 Can you explain the numerical data on a post-box?
  When is the next collection?
  Look at your watch and find out how long it will be to the next
  collection.
5 You need to phone home from a public call box. Do you know the
  dialling code for your area?
6 Look for numbers in unexpected places – lamp-posts, old buildings,
  advertisements – and try to interpret what they mean.

FLOWERS

*Sorting/classifying Number patterns/ideas Patterns
Shape/size Probability/statistics Arranging*

Look at a display of flowers.

How many are red?
How many are yellow?

Are there any flowers that have the same shape?
                              the same pattern?
                              the same number of petals?

How many types of 'daisy' flower can you find?
Which is tallest?

AT THE GARAGE

Capacity/volume Money

The car needs filling with petrol.

How much does the tank hold?
What is the price of petrol: in gallons, in litres?
Can you read the digital displays on the petrol pump?
Which display shows the amount of petrol delivered? How many litres
is that?
How much must be paid for the petrol delivered?

87

LEAVES

*Language/talking Sorting/classification Patterns Area
Length/distance Shape/size*

Make a collection of all the different leaves you can find in the
environment around you.

Ask the children to sort them in different ways and tell you why they have
sorted them in that way. They may sort for colour, size, shape, texture,
smell, length, width, edges (toothed or smooth), etc.
Once sorted, can they find another way? Listen again to their
explanation.

When you get home, ask them to make rubbings of the leaves with a wax
crayon. Notice the vein patterns.

Look at the *area* of each leaf by drawing around it on squared paper and
counting the squares.

Take a piece of cotton and lay it around the edge of a leaf. Trim the
cotton so it is the same length as the leaf's edge. Repeat for several
leaves.
You can now compare the *perimeter* (the distance around the outside) of
the leaves by putting the cotton in order of length. Always let children
estimate first – which leaf will give the longest piece of cotton? This helps
towards understanding perimeter although you don't need to use this
word.

A VISIT TO THE LIBRARY

Language/talking Sorting/classifying Time

Books can be simply classified as fiction or non-fiction, but under these
headings come many more classifications. For instance, the science
section may contain, amongst others, books about biology, chemistry
and physics. The coding and reference systems in a library are important
as they enable us to find specific books quickly. The ability to classify is
an important mathematical skill often needed for problem-solving. Some
of the following activities will help young children develop this skill.

Where will you find a story-book?
　　　　　　　　a book about space?
　　　　　　　　a book about birds?
Is there a code or reference to help you?

What do the letters stand for?

Are the books that are higher up for older children?
Where is the largest book? Can you borrow it?
What date is it today?

Now have your books stamped!
On what day and in which month are your books due back?
How many weeks away is that?

THINGS TO DO ON A SNOWY DAY

Patterns   Temperature   Capacity/volume   Shape/size

Look at footprints and animal tracks and decide which way they are
going.
How many different-sized footprints can you see?

Collect some snow in a container (possibly a plastic measuring basin or
jug from the kitchen).
How much water is left after the snow has melted?

Try to make a snowman that is one metre in height.

Use a thermometer to find out the temperature indoors and then
outdoors.
Compare the temperatures. What is the difference?
Did the temperature outside fall below zero?

When indoors use a calculator to explore ways of getting from a number
above zero to a number below zero.

Did you know that every snowflake is unique? If you examine snow
crystals under a microscope you will find they are all hexagons (six-sided)
and yet all are different in design.
Look at some snowflake designs.

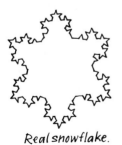

Own design –
from paper.

Real snowflake.

Design your own 'snowflakes' by folding and cutting symmetrical shapes
from circles and hexagons.

89

BRICK PATTERNS

Shape/size Patterns

Look closely at the brick houses, walls and buildings near your school
and home. Notice the different patterns. Sometimes all the long sides of
the bricks show, sometimes the ends show.
Copy the patterns on to a piece of paper.
How many can you find?

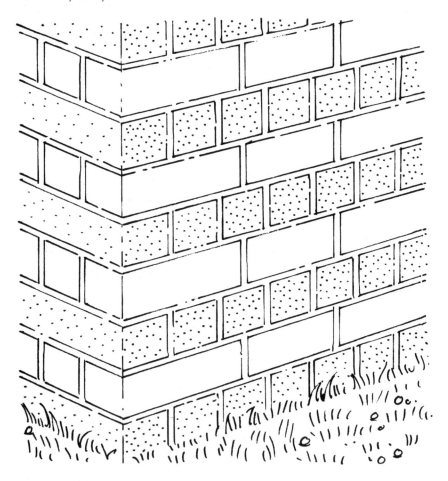

Each pattern has a special name. Can you find out what they are called?
For example, the one illustrated is called 'English bond'.

When shapes fit together like bricks we say they *tessellate*. Other shapes
tessellate too. Can you find any?

Outdoors

7- to 11-year-olds

ESTIMATING STEPS

Length/distance   Estimating/approximating   Ratio
Predicting   Time

As you walk along the road together, consider these questions:

How many steps do you think you will take from here to the corner or to the next lamp-post?
Count and check your estimate.
Are all the lamp-posts the same distance apart?

Is the number of steps the same for each member of the family?
Can you spot a relationship between the number of steps taken by your child and the number you take yourself? By knowing one score, can you predict the other? (These questions are appropriate only for older children and may need to be done with a calculator.)
Try predicting for each other: 'I think you will need between 20 and 25 steps to get from here to the telephone box.'
At what distance does it become very difficult to make a reasonable estimate?

If one person adjusts their step in order to prove the estimating person wrong, this is also mathematically interesting.
If you want to reduce the number of steps taken, how do you adjust the length of your step?

Do you find yourself needing an unexpected number of steps on slopes?
Is it the same for going up as for going down?
Why do you think it is?

SPOT THE SIGNS

Addition   Subtraction   Collecting/using information

On a car journey pick a section of road between one town and the next, or decide on a fixed amount of time going through an urban area.

With or without a calculator ask the children to keep a running total of the value of all the speed signs they see, i.e. 30, 40, 50, 60, 70 – or 10 perhaps in a tunnel. (The 'end of speed restriction' sign – see diagram – counts as 60.)

The aim is to reach a thousand.

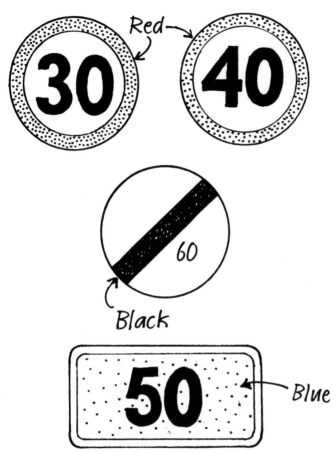

Variation

Give a starting figure of a thousand and do a running total subtraction with the calculator.

The aim is to reach zero.

TEST YOUR MEMORY

Language/talking   Estimating/approximating

This will be a suitable activity if you have tried the one called 'ESTIMATING STEPS' (see page 92).

Two of you go for a walk along a familiar route. One of you must close your eyes and give a running commentary of what you are passing. Alternatively, the person whose eyes are open could ask questions.

93

BEAT THE CLOCK

Time   Length/distance   Speed

This can be done anywhere there is plenty of space.

Set up an obstacle course using anything that comes to hand: clothing, footwear, stools or folding chairs, rugs or cushions, plastic containers. . . . These can be used to jump over, crawl under, balance on the head or hand, jumped with held between the knees or feet, hopped around, etc. Each person taking part has their performance timed in minutes and seconds.
The aim is for each person to have two or three goes and improve their own time.

Variation
Complete the course in the slowest time possible, without stopping at any point.

TAKE

Subtraction   Predicting   Logical thinking

You will need sticks, stones, shells, or any available small objects. You need 30 altogether.

Place them in a pile. Decide who will go first, then take it in turns.
You can pick up any number of objects from 1 to 6,
The winner is the one to pick up the last object.

Try the game with a different starting number and being able to pick up a different number of objects.
Example: 32 objects. You can pick up 1 to 5 objects.

Can you develop a strategy so that you always win, whatever the numbers?

I'M THINKING OF TWO NUMBERS

Mental calculation   Multiplication   Addition

This is a game that can be played anywhere, by any number of people. At first the numbers should be kept small, one under five, the other under ten.

One person begins by saying:
   'I'm thinking of two numbers;
   In total they add to . . .
   When multiplied together they produce . . .
   What two numbers am I thinking of?'

Keep everyone thinking by keeping the same total and changing the product. For example, the total is 10, the product is 25, followed by the total is 10, the product is 16.
Or try keeping the same product and varying the total. For example, the total is 7, the product is 12, followed by the total is 8, the product is still 12.

HOW FAR CAN YOU JUMP?

Estimating/approximating   Length/distance
Sorting/classifying

This is for adults who wish to lie quietly whilst the children are active. You need an empty stretch of beach or grass.
Clearly mark a long starting line.
Each child taking part lies down, feet against the starting line, and their height is marked.
Using a standing jump only, the children try to jump their own height.
They can do this using both feet to both feet
                         right foot to both feet
                         left foot to both feet

Ask:
   If they jumped less than their height – was it about halfway, more than halfway or less than halfway?
   If they jumped more than their height – can they estimate how much more, using half heights, or even double the height?
   Can they jump the height of a person taller than themselves?
   Can they jump twice the height of a shorter person in the group?

PAVING STONES
(a strategy game for 2 players, 3 at a pinch)

Mental calculation   Logical thinking

It can be played anywhere, e.g. while travelling or sitting in a waiting room.

Players need one coloured pencil each, a piece of paper folded or cut into a triangle, something on which to rest the paper.

Mark the paper all over with crazy paving.
Players then take it in turns to colour in one of the paving stones.
The winner of the game is the first player to make a continuous path which touches all three edges of the triangle.

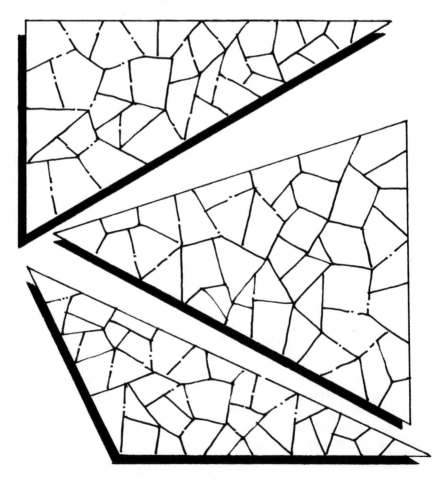

IN THE PARK

Length/distance Predicting Shape/size

How far do you think you can throw a ball?

You probably won't have a measuring-tape at the park, so measure in paces.
Study bouncing balls. Which bounces highest?
Does the ball with the highest bounce, bounce the most number of times before coming to a full stop?
From what height do you need to drop your ball for its first bounce to come up to your knee?

Watch a ball being thrown from person to person.
If the ball drew a line as it travelled, what would the line look like?
How does this line change as the ball is made to travel faster through the air?
Imagine the line drawn by a ball bouncing along the ground.
What about swings and see-saws? Can you imagine their lines? Try to draw some of these when you get home.
When watching television, notice the lines of high-jumpers, long-jumpers, pole-vaulters and others.

WHERE ARE THEY?

Language/talking

This is an activity that can be done almost anywhere.

One person selects a stationary person and describes their position. For example:
The person is to the left of, to the right of, in front of, behind, next to . . .

If players don't get the correct person, they're allowed to ask questions. The questions should also be in terms of position only. Questions about clothing, for example, are not acceptable.
Any building or stationary object can also be used.

NUMBER-NAMES

Language/talking   Investigating

Discuss words or parts of words that mean a certain number. For
example:

    mono   meaning 1

    bi   meaning 2 (What other words tell you there are 2?)

    tri   meaning 3

    quad   meaning 4

Look for objects which use this system in their name. Bicycle and tricycle
are easy; can you find any others?

How many of these objects can you see on one journey?

How about words that start with letters that mean a number greater than
4? Can you find the words for 5, 6, 7, 8, 9, 10, 100 and 1000?

How many examples of objects with 'number-names' can you find?

OUT 'DOORS'

Logical thinking   Addition   Number patterns/ideas
Place-value

You can invent this type of problem as you walk along the road. Keep the numbers simple to start with. Older children will be able to handle bigger numbers, especially with a calculator.

Look at the numbers on the houses on both sides of the road. What do you notice? (Children will need to understand that the even numbers are on one side and the odd numbers on the other.)
Choose two houses next to each other and ask children to add up the numbers on the doors.

Now for the problems:
   The numbers on two houses next to each other add up to 8. What are the numbers on the doors?
   If the numbers add up to 26, what are the numbers on the doors?
   How about 106? Is there a quick way to work this out?

Some houses in the road do not have their number showing. Can you work out what number they are?

99

HOW ACCURATE ARE ROAD SIGNS?

Decimals   Addition   Subtraction   Fractions   Calculator use

This can be done on any car journey. You will need pencil and paper, and a calculator.

Note down the first road sign with your destination and number of miles to travel on it. At the same time ask the driver to give you the milometer reading.

Example: The sign – London 112. The milometer reading – 28.4. At the next sign, note the miles and the milometer reading. Check the accuracy of the road sign against the milometer.

Example: If the next sign reads, London 98, and the milometer reads 42, then you can do two subtractions to check the road sign's accuracy. I.e.

$$112 - 98 = 14$$
$$42 - 28.4 = 14 \text{ (approx.)}$$

This can be a more interesting activity when road signs on country lanes are read. Everyone has to decide whether the place named really was, say, 4 miles further along the narrow, winding road. When you've all made your guesses, check the milometer reading.

Quite often a signpost will read something like 'Purton 3¾'. You must convert this fraction to a decimal to check the milometer reading. (You will find a calculator useful when adding 3.75 to the milometer reading.) Remember – most milometers only have one decimal place.

SHADOWS

Length/distance   Estimating/approximating
Angle/direction   Time

How confident are you about estimating the height of a lamp-post or a tree?
Try it, and then check your estimate with this simple method:

Put a long stick in the ground in your garden. Measure its height above the ground. Then measure the length of its shadow.
Keep measuring the shadow every half-hour. What do you notice?
Is there a time when the stick and the shadow measure the same or when the shadow is twice the length of the stick?
Can you use this idea in any way to help you check your estimate of the height of the lamp-post or tree?

A VERY SPECIAL PATTERN

Patterns

Can you see how this series of numbers is developing?

1   1   2   3   5   8   13   21   34   55

Each number is the sum of the two which go before. This is known as the *Fibonacci series*, after the mathematician who first used it – a very long time ago. (Fibonacci lived from 1170 to 1250.)

It's a specially interesting series because it can be found in many aspects of the living world.

Look at the spirals on a pine-cone or a dahlia flower. If you can count them (and it's not easy) you'll find the number of seed cases or petals in each spiral follows the Fibonacci series.

You may also find a flower which grows like the one in the diagram below. See how the series develops:

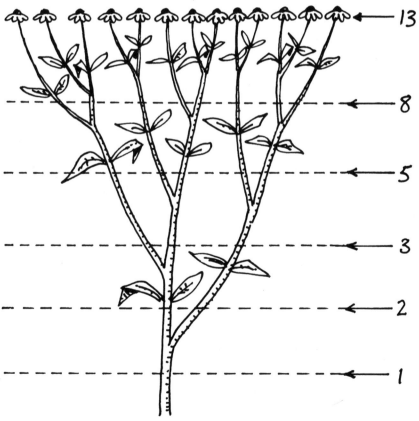

BEST BUY

Money   Estimating/approximating   Capacity/volume   Weight

You will need a calculator.

At the supermarket, look for two or three sizes of the same product.
Guess which is the best buy and then use a calculator to check whether
you are right.
Is it always cheaper to buy the largest size?
Look at packets that are sold by weight as well as containers that are
sold by capacity. Can you work out how much could be saved by
purchasing these 'best buys'?
What would the saving be in one month,
in one year?

Also in the supermarket, can your child keep a running total, in their
head, of how much you have spent by rounding up or down to the
nearest £? For example:

   87p – £1   £2.30 – £2
   35p – £0   £1.45 – £1

How accurate is this method?
Can they devise a more accurate one of their own?

CAR NUMBERS

Arranging   Addition   Number patterns/ideas

Car registration numbers can be a great source of mathematical ideas to
brighten journeys by road. Here are some ideas – there are many more.

1 How many ways can you rearrange your car number, using just the
   numbers? How many ways if you use the letters as well?
2 Choose a number, say 15, and see who can spot a car whose
   registration number adds up to this total, e.g. E294 VYB.
3 Can you spot a car with all even numbers in its registration number?
   (Or all odd numbers, all prime numbers, all less than 5, all more than
   7, etc.)
4 Who can spot the highest/lowest total in one car number for the
   journey?
5 Can you work out the age of the car from its registration letter?
6 Can you find the place the car was registered from the AA list? Use a
   road map to find how far the car is from its home.

In the Car or on a Bus

DESTINATION CODES

Addition   Multiplication   Number patterns/ideas

This is a good game for any car or coach journey, but you need to do some preparation before the journey begins.

Write the letters of the alphabet on a piece of paper. Give each letter a value.
For example:
   A = 1, B = 2, C = 3, etc.

Write down the names of the home town and the destination town and record their number values.
For example:
   Bristol would be 2 + 18 + 9 + 19 + 20 + 15 + 12 – a total of 95.
   London's value would be 74.

During the journey write down the name of every town or village passed through and calculate its value. (On a motorway, count the towns named at each junction.)
The aim is to find the town with the largest value, excluding the home or destination town.

Variations
1 Find the town with the smallest value.
2 Find a town which has the same value, or closest value, to the home or destination town.
3 Find as many towns as possible whose values lie between the home and destination towns.
4 Value the letter A at 26, B at 25, etc.
5 Value the letters using odd or even numbers.
6 Value the letters using multiples of 3, 5, etc.
For example:
   A = 3, B = 6, C = 9, etc.

NUMBER PLATES

Addition   Multiplication   Division

Win points by spotting a car number which is:

1 a multiple of ten (1 point)
2 a multiple of five (2 points)
3 a multiple of three (3 points)

Quickly add the digits together.
Which seems to be the most frequent total?
Any ideas why?

ETA (Estimated Time of Arrival)

Time   Predicting

At the start of the journey, everyone estimates the time of arrival.
Whose estimate was closest?

GAUGING RESULTS

Capacity/volume   Speed   Length/distance   Predicting

Encourage your child to look at instruments and gauges. Talk about them and how they work. For example,

The petrol pump: 'How many litres of petrol did we get and what did it cost?'
The fuel gauge: 'Is the petrol tank more or less than half full?'
The speedometer: 'How fast are we going?'
The odometer: 'How many kilometres still to go?'
'How many kilometres from town A to town B?'

ROAD SIGNS

Logical thinking   Language/talking

Think about road signs. Some are quite small. Others are very large. Why do you think this is? Do the largest signs appear on a certain type of road?

On country roads, you may not see the signpost for which you are looking until you are right up to the junction. On motorways, you are told several times about the turn-offs ahead. Why do you think this is?

Look at any words written on the road surface? Why do you think the letters are so long?
Look at this sign:

What do you think it is telling you? Why are there no words on the sign?
What signs do we use in mathematics?
Why do we use them?
Are they easy to understand?

MORE ON SHADOWS

*Time Angle/direction Investigating
Collecting/using information*

Think back to the activity called 'SHADOWS' (see page 100), or do it now.

Did you notice that the length of the stick's shadow and the lengths of all other shadows, including those of the lamp-post and tree, changed?
Did you notice also how the *position* of the shadow was changing?
Did you notice the direction in which it moved?
Look out for a sundial and see if you can find out how it works. You could even try to make a simple one in your garden.

When you experimented with the stick, at what time of day was the shadow shortest?
At that time (if you live in the Northern Hemisphere) the sun must have been directly south because of the sun's movement from east to west.

Early morning.

Midday.

Evening.

E   W

There's another way of finding out in which direction south lies. You will need a watch or clock. It must be the traditional analogue kind, not digital.

Go outside and put the watch on the ground, pointing the short hand towards the sun. Now look at the angle (clockwise) between 12 and the short hand. Imagine a line drawn from the centre of the watch-face which would halve this angle. That line points to the south.

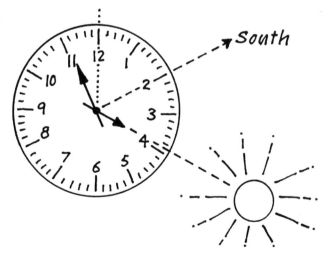

Think of what would happen at midday.
The sun is in the south at midday.
If you are outdoors or in the car at midday, look for the sun. Work out in which direction north, east and west lie. Notice the way in which the sun visors in your car can be angled.

TIMBER!

Estimating/approximating Length/distance Multiplication

These activities are a challenge for adults and children alike. The aim is to estimate the height of a tall tree. The measurements are all made in paces. When you get home you can measure the length of a pace and then work out the height of the tree in metres and centimetres.

Method 1

When you're out, get the family to stand around a tall tree. Ask them to move to the place where, if the tree should fall, they think the tip would fall at their feet. Mark the position.
If they're right, the distance from here to the base of the tree is equal to the height of the tree.

Method 2

This method can be used to find the height of anything tall. You can do it unaided, provided there is a clear, level area of ground in front or behind the object to be measured.
Stand with your back against the tree and pace away from it, keeping a count of the number of paces. Do this until it is possible to look back through your legs and just see the tip of the tree.
Indian scouts reckoned the distance in paces was about equal to the height of the tree.

Method 3

This is probably the most accurate method. Use it to check the results of methods 1 and 2. You will need two people and a pencil.

Ask the other person to stand in front of the tree, then estimate how many of that person it would take to reach the top of the tree. Use the pencil to do this (see diagram).

If the person's height is not known, then a rough guide is that someone's height is about the same as two even paces.

So, if the estimate is four-and-a-half times that person, the height of the tree would be about nine paces.

BY THE STREAM

Time Predicting Shape/size Language/talking

One person can do this alone but it is more fun with two. You will need a selection of twigs.

Stand on a bridge if there is one – that way you have a better view. Otherwise, the bank will have to do.

First you need to agree on a finishing-post for your twigs, which are going to float downstream.

Then each of you must estimate the number of seconds you think your twig will take to reach the finishing-post.

Then drop your twigs in and count together.

There are two ways of scoring points: either by having the fastest twig or by making the best estimate. Score one point for each. That way, you could both be winners!

Now do it again. . . .

When you drop something in the water, watch the ripples.
Describe what happens.
Ripples form *concentric circles* (circles inside circles). Can you describe how these are different from a circular spiral?

PATTERN OUT-OF-DOORS

Shape/size Symmetry Patterns Language/talking

Look for patterns when you are out and about.
Can you find patterns on walls, fences, trees?
What makes a particular set of shapes into a pattern?
Strange, isn't it? We all feel we know what pattern is, but could we explain how we know?

To work out what pattern is, try thinking about a heap of bricks tipped from a truck, and a brick wall. We could say that the heap of bricks shows a random pattern. What does the wall show?

Think of other designs and try to explain to each other why they are patterns.
Are these words helpful: order, sequence, symmetry, repetition?
Are these words clues for the detection of pattern?
Look at the shadow cast by a wire fence. Do the shapes in the pattern of the shadow match the shapes in the pattern of the fence itself?
How often to squares, rectangles, triangles, circles, etc., appear in the patterns you see out-of-doors?

Which patterns are artificial and which are natural?
Are the shapes of natural things different from those of artificial objects?

HOW FAR?

Length/distance   Estimating/approximating   Average

Make paper aeroplanes – of several different designs, if you can.
Estimate which will fly the farthest and how far, in metres, it will go. Test them all to see if you were right.
Choose the best design. Try ten flights and measure the distance flown each time. Can you work out the average length of flight?

Make a slope and choose several toy cars. Can you estimate which car will travel the farthest down the slope?
Does the car that goes the fastest go the farthest?
What difference does it make when you alter the angle of the slope?
Can you find the *optimum angle* (the angle that produces the longest distance)?

What else could you investigate in this way?

WHEELS

Length/distance   Estimating/approximating

Children are probably familiar with a 'trundle wheel' (a wheel that measures 1 metre in each revolution) through using them at school. Their bicycle wheels can be used in the same way, but they will not measure a metre.

What is the distance covered for one revolution (circumference) of the bicycle wheel?
Use the tyre valve as a marker. Estimate how many times the wheel will revolve:

   to cross the road
   to the next lamp-post
   to the next turning, etc.

Measure the diameter of the wheel (the distance from one side to the other through the hub).
Is there any connection between the diameter of the wheel and the distance covered in one revolution?
If there is, is the same true for their friend's or brother's or sister's bicycles that have different size wheels?

111

LOGOS

Length/distance   Symmetry   Sorting/classifying   Shape/size
Angle/direction

While you are out the whole family can look for logos that advertise or represent different shops, companies, services or cars.
Perhaps the best-known logo is for British Rail.

Back at home, you can copy and enlarge these logos, perhaps using squared paper. Children will need to observe closely and measure accurately to do it well.
Next, classify the logos into symmetrical and non-symmetrical. Use a mirror to find the lines of symmetry.

Some logos have *rotational symmetry*. This means that, during a rotation, the image fits exactly on to the original object.

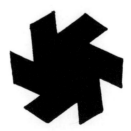

Look for other examples like this. Keep a collection of logos in a scrapbook.

Children could design a logo for a particular purpose, for example:
  for a club they belong to
  for their favourite car or the company that makes their favourite toys or games
  for their bicycle
  for their school or family

GARDENING

Addition   Multiplication   Division   Length/distance   Time
Patterns   Money

Planning a garden, buying plants and seeds and then planting them involves a lot of maths. Try to bring this out when the children want to help. Here are some ideas; you're bound to think of more once you start.

Flowers
Look at the heights the plants will grow to on the seed packet. Arrange them with the tallest at the back of the flower bed and the shortest at the front.
Look at the colours. How can the plants be arranged to make the best display?
Look at the time of year the plants will flower. Can you arrange for there to be flowers blooming from March to October, or even all the year round?
Look at the recommended distance between plants. How many will be needed?
Make a calendar of when to sow your chosen varieties, using the information on the seed packet.

Vegetables
Measure the width of the plot. Work out how many of each type of vegetable plant you will need for a row, depending on the distance apart they must be planted.
How deep must the seeds, plants or tubers be put?
How can you sow the seeds so that you have a succession of vegetables throughout the year?
Can you devise a simple way to measure the distance between plants when you plant them?

BUZZ AND FUZZBUZZ

Multiplication

This is a fun way to practise multiplication tables during a journey or whenever you have a free moment. Any number of people can play.

Buzz

To play Buzz, choose a table.
Start to count, but each time you reach a number which comes into the table you say 'buzz'.
Example:
For the 4× table
1 2 3 buzz 5 6 7 buzz 9 10 11 buzz 13 14 etc.
Carry on until someone claims you've made a mistake. If they're right, they have a turn.

Fuzzbuzz

To play Fuzzbuzz, choose two tables. Replace the numbers in one with 'fuzz' and those in the other with 'buzz'.
Example:
For the 3× and 5× tables call out:
1 2 fuzz 4 buzz fuzz 7 8 fuzz buzz 11 fuzz 13 14 fuzzbuzz 16 17 fuzz 19 buzz fuzz 22 23 etc.

Support for change

Support for the developments that have taken place in mathematics has been growing. Not only do teachers, parents and teacher-trainers recognise this more enlightened view of mathematics but so does the government. In 1978 the government established a committee of enquiry into the teaching and learning of mathematics in primary and secondary schools in England and Wales. The committee, chaired by Dr W. H. Cockcroft, was asked to conduct a survey and prepare a report 'with particular regard to the mathematics required in further and higher education, employment and adult life, and to make recommendations'.

The committee's report, endorsed by Sir Keith Joseph, then Secretary of State for Education and Science, was published in 1982. Its title is *Mathematics Counts*, but it is usually referred to as the 'Cockcroft Report'. Its recommendations have had a profound effect on the development of mathematics education, particularly in the last five years, because they challenge many long-held assumptions about the teaching and learning of mathematics.

In particular the Report called attention to a need for primary schools to foster the development of general strategies for problem-solving and investigation. It recommended that children be given opportunities to become familiar with such processes as:

- making a graphical or diagrammatic representation, and observing pattern in results;
- making a conjecture and discovering whether or why the conjecture is correct or not;
- learning how to solve a problem by looking at a simpler related problem;
- developing persistence in exploring a problem;
- setting up an experiment and recording the possibilities arising from it.   115

It also had the following to say about learning mathematics:

> A premature start on formal, written arithmetic is likely to delay progress rather than hasten it. . . .
>
> It has been pointed out to us that, albeit with the best intentions, some parents can exert undesirable pressure on teachers to introduce written recording of mathematics especially 'sums' at too early a stage because they believe that the written record is a necessary sign of the child's progress. . . .
>
> In our view it is right that primary teachers should allow children to use calculators for appropriate purposes. . . .
>
> . . . emphasis on arithmetical skills does not of itself lead to ability to make use of these skills in practical situations. . . .
>
> To speak of logic in connection with young children may surprise some people. In games and puzzles, moves often have to be made according to rules and finding the best moves involves logical thought. . . .
>
> All children need experience of practical work which is directly related to the activities of everyday life, including shopping, travel, model-making and the planning of school activities. Children cannot be expected to be able to make use of their mathematics in everyday situations unless they have opportunity to experience these situations for themselves.

These Cockcroft recommendations offer support to the many educators who believe, as we do, that in order to educate our children to meet the unpredictable challenges of the year 2000 and beyond we need to present children with unfamiliar problems in the form of activities and thinking games like those in this book. Experience shows that children are capable of learning such things as logical reasoning, awareness of mathematical pattern, recognising relationships and the sophisticated use of mathematical language. Through activity young learners can be encouraged to become alert and open-minded in their approach to maths, and this will serve them well in the future.

We hope that you will enjoy working with your child on some of these activities. If you are keen to find out more, ask for mathematical books at your local library. To help you in this, a selection of reading books is included on the next page. They are suitable for primary age children and offer opportunities for discussion about mathematical ideas. A short list of suggestions for presents (generally under £5) which have a mathematical bias is also offered.

Bibliography

Books for younger children

AHLBERG, A. and J. *Each peach pear plum* Viking Kestrel, 1984; Armada, 1980.

ARMITAGE, R. and D. *The lighthouse keeper's catastrophe* Penguin, 1988.

BAYLEY, N. *The patchwork cat* J. Cape, 1981; Penguin, 1984. op.

BERENSTAIN, S. and J. *Bears in the night* Collins, 1981.

BIRMINGHAM, D. *'M' is for mirror* Tarquin, 1988.

BUCKNALL, C. *One bear all alone* Macmillan, 1985; pb, 1987.

BURNINGHAM, J. *The shopping basket* J. Cape, 1980; Armada, 1983.

CARLE, E. *The bad-tempered ladybird* H. Hamilton, 1978; Penguin, 1982.

CARLE, E. *The very hungry caterpillar* H. Hamilton, 1970; Penguin, 1974.

COOKE, B. *The little fish that got away* Scholastic, 1965. op.

GREENE, G. *The little train* Penguin, 1977.

GARLAND, S. *Going shopping* Bodley Head, 1982; Penguin, 1985.

GRETZ, S. *Teddybears one to ten* Collins, 1973.

HEIDE, F. P. *The shrinking of Treehorn* Penguin, 1975.

HILL, E. *Spot learns to count: colouring book* Heinemann, 1984.

HODGSON, J. *Pointers* Ginn, 1987.

HUTCHINS, P. *The doorbell rang* Bodley Head, 1986; Penguin, 1988.

HUTCHINS, P. *Rosie's walk* Bodley Head, 1968; Penguin, 1970.

HUTCHINS, P. *Titch* Bodley Head, 1972; Penguin, 1974.

KAHN, P. *Ten little bears counting book* Random House, 1983.

KERR, J. *The tiger who came to tea* Collins, 1972; pb, 1973.

MCKEE, D. *Elmer* Dobson, 1968; Pan, 1973.

ROFFEY, M. *Farming with numbers: a picture book* Bodley Head, 1972. op.

ROSS, T. *Towser and Sadie's birthday* Anderson, 1984; Armada, 1985.

ROSS, T. *Towser and the magic apple* Armada, 1986.

ROSS, T. *Towser and the terrible thing* Anderson, 1984; Armada, 1986.

ROSS, T. *Towser and the water rats* Anderson, 1984; Armada, 1985.
SENDAK, M. *Where the wild things are* Bodley Head, 1967; Penguin, 1970.
SHANNON, G. *The surprise* J. MacRae, 1984; Macmillan, 1985.
WALTER, M. *The magic mirror book* Hippo, 1984.
WALTER, M. *The second magic mirror book* Hippo, 1984.
WILDSMITH, B. *Professor Noah's spaceship* OUP, 1980; pb, 1985.
YEOMAN, J. *Sixes and sevens* Macmillan, 1988.

Books for older children

ABDELNOOR, R. E. J. (ED.) *A mathematical dictionary* Wheaton, 1979.
ANDERSON, S. *Dragon-hunting book of Knightly puzzles* Pan, 1987.
BALL, J. *Think of a number* BBC, 1979. op.
BROWN, J. *Flat Stanley* Methuen, 1977.
EMMET, E. *The Puffin book of brain teasers* Penguin, 1987.
FARRIS, D. *Amazing* Unwin, 1982. op.
GALLO, G. *The lazy beaver* Collins, 1983; Pan, 1985.
HUGHES, T. *The iron man: a story in five nights* Faber, 1985.
JUSTER, N. *The phantom tollbooth* Collins, 1974.
KING-SMITH, D. *The fox-busters* Gollancz, 1978; Penguin, 1980.
LANE, M. *Operation hedgehog* Methuen/Walker, 1981; Magnet, 1984.
MESSITER, I. *The incredible quizbook* Unwin, 1983.
OAKLEY, G. *The church mice chronicles* Macmillan, 1986.
PELHAM, D. *The Penguin book of kites* Penguin, 1976.

Suggestions for presents

Suggestions for presents – generally around £5

Type/name of game	Nature of game	Manufacturer or retailer
Wooden puzzles	range of mathematical puzzles available	Bristol Council for Disabled, 75 Whiteladies Road, Bristol, BS8 2NT
Continuo or Triangular Continuo	a game of counting and colour matching	Hiron
Making Shapes		Tarquin Publications
Play and Learn Calendar Ladybird Clock	manual alteration games	Uniset Ladybird
Playing cards	regular games like 'Snap', as well as specialist packs for 'Happy Families' etc.	toyshops and stationers everywhere

Construct-O-Straw *Lego* *Meccano* *Plasticine*	construction games	all available from ELCs, and toyshops
Dice *Solitaire* *Snakes & Ladders* *Ludo* *Chinese Chequers* *Chess* *Draughts* *Jigsaw puzzles*	all games of adding and counting as well as investigation	all available at ELCs, Mothercare and toyshops in general
Spirograph	game of drawing patterns	toyshops
Skirrid	a type of chequers	toyshops
Connect Four	3-dimensional noughts & crosses	toyshops
Calculators	simple ones are available for children eg Texas T1 1103	W.H. Smith, Argos etc.
Jeu de Miroir	a mirror game, symmetry and patterns along axes	Nottingham Educational Supplies, 17 Ludlow Hill, West Bridgford, Nottingham, NG2 6HD

Index

In the home *4–8-year-olds*

In the home *7–11-year-olds*

Addition
Total 100 47; Quick on the draw 50; Password 51; Addsnap 52; Elevenses 52; Guess what the computer is doing 54; Make 15 56; Rich for a day! 58; What day did it happen? 58; Letter puzzles 60; First to 20 61; Dead on target 64; Hit the target 65; Faulty keys 66; Perfect numbers 66; Pocket-money 66; Sleepy heads 69; TV addicts 71

Angle/direction
Square corners 62; How far will it open? 69

Area
Exploring numbers 48; Plots 61; Tangram 67; Do-it-yourself 68

Arranging
Holiday packing 48

Average
They say . . . (1) 52

Calculator use
Rich for a day! 58; Dead on target 64; Faulty keys 66; TV addicts 71

Capacity/volume
Make a container 50; Do-it-yourself 68

Collecting/using information
What day did it happen? 58; Weather countdown 68; Body maths 70

Counting
Using a calendar 49

Decimals/percentages
Sleepy heads 69

Division
Sweets 47; Quick on the draw 50; Password 51; Guess what the computer is doing 54; What day did it happen? 58; Plots 61; Dead on target 64; Hit the target 65; Faulty keys 66; Perfect numbers 66; Thunder clap 72

Estimating/approximating
Exploring numbers 48; Make a container 50; Largest product 56; Rich for a day! 58; Square corners 62; Pocket-money 66; Do-it-yourself 68; Body maths 70

Fractions
Dead on target 64; Tangram 67; Sleepy heads 69

Investigating
Sweets 47; Total 100 47; Crossing the river 51; Largest product 56; Hidden shapes 63; Perfect numbers 66; Tangram 67

Language/talking
Exploring numbers 48

Length/distance
They say . . . (1) 52; Do-it-yourself 68; Body maths 70

Logical thinking
Square puzzle 46; Crossing the river 51; Make 15 56; Consecutive numbers 57; Holiday choice 59; Letter puzzles 60; First to 20 61; Plots 61; They say . . . (2) 62

Mental calculation
Using a calendar 49

Outdoors 4–8-year-olds

Outdoors *7–11-year-olds*